RUSSIAN IMPERIAL MILITARY DOCTRINE AND EDUCATION, 1832–1914

Recent Titles in
Contributions in Military Studies

Feeding the Bear: American Aid to the Soviet Union, 1941–1945
Hubert P. van Tuyll

Military Planning for the Defense of the United Kingdom, 1814–1870
Michael Stephen Partridge

The Hollow Army: How the U.S. Army Is Oversold and Undermanned
William Darryl Henderson

Reevaluating Major Naval Combatants of World War II
James J. Sadkovich, editor

The Culture of War: Invention and Early Development
Richard A. Gabriel

Prisoners, Diplomats, and the Great War: A Study in the Diplomacy of Captivity
Richard Berry Speed III

Military Crisis Management: U.S. Intervention in the Dominican Republic, 1965
Herbert G. Schoonmaker

The Persian Gulf War: Lessons for Strategy, Law, and Diplomacy
Christopher C. Joyner, editor

Where Eagles Land: Planning and Development of U.S. Army Airfields, 1910–1941
Jerold E. Brown

First Strike Stability: Deterrence after Containment
Stephen J. Cimbala

Legitimacy and Commitment in the Military
Thomas C. Wyatt and Reuven Gal, editors

The Army's Nuclear Power Program: The Evolution of a Support Agency
Lawrence H. Suid

RUSSIAN IMPERIAL MILITARY DOCTRINE AND EDUCATION, 1832–1914

Carl Van Dyke

CONTRIBUTIONS IN MILITARY STUDIES,
NUMBER 105
John Erickson, *Series Adviser*

Greenwood Press
NEW YORK • WESTPORT, CONNECTICUT • LONDON

Library of Congress Cataloging-in-Publication Data

Van Dyke, Carl.
Russian imperial military doctrine and education, 1832-1914 / Carl Van Dyke.
p. cm. – (Contributions in military studies, ISSN 0833-6884 ; no. 105)
Includes bibliographical references and index.
ISBN 0-313-27249-2 (lib. bdg. : alk. paper)
1. Soviet Union–Armed Forces–History–19th century. 2. Soviet Union–Armed Forces–History–20th century. 3. Strategy.
4. Military education–Soviet Union–History–19th century.
5. Military education–Soviet Union–History–20th century.
6. Military art and science–Soviet Union–History–19th century.
7. Military art and science–Soviet Union–History–20th century.
I. Title II. Series.
UA770.V35 1990
355'.007'047–dc20 90-36062

British Library Cataloguing in Publication Data is available.

Library of Congress Catalog Card Number: 90-36062
ISBN: 0-313-27249-2
ISSN: 0833-6884

First published in 1990

Greenwood Press, 88 Post Road West, Wesport, CT 06881
An imprint of Greenwood Publishing Group, Inc.

Printed in the United States of America

The paper used in this book complies with the Permanent Paper Standard issued by the National Information Standards Organization (Z39.48-1984).

10 9 8 7 6 5 4 3 2 1

The study of the history of military thought helps us seek out and discover answers to the most important questions of military theory. Without knowledge of the historical roots and without knowing the circumstances from which one or another theoretical concept was formed one would, at times, be hard-pressed to predict its correct future development.

Colonel-General V. N. Lobov

Contents

Illustrations ix

Acknowledgments xi

Introduction: Military Education and the Operational Significance of History in Russian Imperial Military Doctrine xiii

1. Innovation: The Foundation of the Nicholas Academy of the General Staff under A. H. Jomini and Nicholas I, 1832–1847 1

2. Revitalization: The Academy as a Military-Scientific Forum in an Era of Reform, 1847–1856 33

3. Transformation: Turning Theory into Practice in the Miliutin Era, 1856–1878 49

4. Disputation: Military Science in the Age of Scientific Positivism and Russian Nationalism, 1878–1904 91

5. Modernization: The Impact of Technology on Military Science and the Search for a Unified Military Doctrine, 1904–1914 131

Epilogue 161

Bibliography 165

Index 181

Illustrations

MAPS

Plan for the Russo-Turkish War of 1877–1878 in the Balkan Theater 81
Russo-Japanese War of 1904–1905 133

FIGURES

1. Adjutant-General Baron Jomini 2
2. Adjutant-General Sukhozanet 7
3. Lieutenant-General Baron Zeddeler 9
4. Lieutenant-General Stefan 36
5. Adjutant-General Count Miliutin 52
6. Components of Military Art 73
7. Adjutant-General Dragomirov 95

Acknowledgments

Throughout the duration of the research which contributed to this study I have relied extensively on the tangible and intangible support of many people, all of whom merit my deepest respect. My thanks go to John Erickson and Kathy Brown at the Department of Defence Studies, University of Edinburgh, who provided guidance toward this book's completion. Also deserving special note are colleagues at the department who, over the last several years, have pointed out fallacious arguments and recommended valuable bibliographic materials: Research fellows Lieutenant-Colonel William O'Malley, Lieutenant-Colonel Andrew Hulse, and Lieutenant-Colonel Joe Lahnstein of the U.S. army, as well as Christopher Bellamy and Alaric Searle.

Reference staff at the following libraries and archival collections also deserve special commendation for providing access and permission to use much of the primary material referred to in this study: the National Library of Scotland and the University of Edinburgh Library in Scotland, the British Library in London, the Lenin State Library in Moscow, the Ministry of Defense of the French Republic—*Service Historique de l'Armée de Terre* at Vincennes, and the Hoover Institution on War, Revolution, and Peace at Stanford. Valuable service in translating particularly difficult Russian quotations came from Ted Barrall. I am also grateful to Bruce Menning at the Soviet Army Studies Office, Fort Leavenworth, Kans., for allowing me to consult his comprehensive study on the Russian imperial army prior to its publication.

Although the practice of history can never attain a truly comprehensive or definitive conclusion, it is hoped that the reader will find that this study has achieved, at the very least, a semblance of equilibrium. All transliterations from the Russian to Latin alphabets were done according to the Library of Congress system and are based on the post-October Revolution reform of the Russian alphabet. In the case of commonly Anglicized proper names (i.e., Moscow, Nicholas I) the Anglicized spelling has been retained. Any textual inadequacies which may exist within this work are entirely the responsibility of the author.

Introduction: Military Education and the Operational Significance of History in Russian Imperial Military Doctrine

The defeat of Napoleon's reconstituted *Grand Armée* on the fields of Waterloo in June 1815 concluded some thirty years of war marked by the introduction of massive conscript armies capable of maneuver and destruction unprecedented in the history of European warfare. Although the Congress of Vienna was to set the course of European politics once more upon traditional diplomatic norms, the legacy of Napoleon's brilliant military career continued to preoccupy the minds of great statesmen, literary figures, and military professionals. Throughout the remainder of the nineteenth century and well into the twentieth, Russian and other European military theorists marveled at the Napoleonic ideal—the individual genius of the commander—and attempted to make this genius the main component in strategic theory. Emphasis was placed on the ability to bring together both superior material and moral military force against an opponent at a decisive location and time.

In the previous century the warring profession had been seen as a relatively prosaic, geometrical affair. Contemporary theories assumed that the commander, demanding complete subordination of troops and officers to his will, possessed a transcendental mind which could see through the incidental chaff in the circumstances of battle to comprehend a strategic universal or certainty of action in the form of normative principles. However, for military theorists and historians coming to terms with the wanton violence and chaos of modern warfare, the study of Napoleon's poetic and destructive genius brought a new perspective to the theory of strategy. It had become the personal expression of a commander who was to make the best of ever-evolving circumstances through the execution of sophisticated judgment:

> If we pursue the demands that war makes on those who practice it we come to the region dominated by the powers of the intellect. War is the realm of uncertainty, three quarters of the factors on which action in war is based are wrapped in a fog of greater or lesser uncertainty. A sensitive and discriminating judgment is called for; a skilled intelligence to scent out the truth.[1]

This recognition of infinite causality in warfare did not deny the utility of a commander's strategy in determining the conduct of his forces in the field. Instead of relying on objective and static principles for guidance, the commander, faced with time and information constraints, was forced to fall back on contingencies of action provided by a complex evaluation of current circumstances with past trends in the evolution of warfare.

Divested of its former mystery the concept of military art and its expression in the actions of the commander was now understood to consist of two personality traits: the ability to instantly and accurately recognize the salient patterns of events in their temporal and spatial significance (a talent called *coup d'oeil,* glance of the eye, and originally attributed to the cavalry officer) in conjunction with the determination to dedicate one's material and moral forces behind the decision taken. In its emphasis on resolute action, be it offensive or defensive, this new cognitive strategy resembled the old, positive strategy in its reliance on self-confidence and single-mindedness in the heat of battle. However, in this case these traits were less a concession to a priori principles and more a comprehension that there was a point where sensitivity to emerging circumstances would increasingly enervate disciplined coordination of superior force against the opponent.

In light of the growing complexity of conducting warfare brought about by the increasing size and technological sophistication of European armies the traditional reliance on the haphazard leadership skills of selected commanders was no longer judged sufficient. Higher military-educational institutions were established in St. Petersburg, Berlin, and Paris to nurture future commanders and provide technical training to an expanding officer corps responsible for the administration of the army's needs. Synonymous with this new institutional emphasis on education was new philosophical fervor over the development of an applicable theory of military art, the most noteworthy intellectual contribution coming from the *Militärgesellschaft* tradition of Gerhard Johann von Scharnhorst and Carl von Clausewitz and the military-theoretical works of Antoine-Henri Jomini.[2]

Contrary to the positive doctrines of the past century the purpose in developing a new theory of military art was to encourage the talented commander to engage in critical investigation of past military experience in order to directly exercise his powers of judgment over circumstance, and ultimately to bolster his self-confidence in battle. Because this analytical exercise was conducted on the specific and empirical basis of military history it retained an a posteriori quality upon which the commander could form subjective theories by which to aid his execution of judgment. In this way these theoretical generalizations could subsequently be compared and evaluated according to the standard of empirical, historical fact from other military experiences. The critical analysis of military history had the added advantage of focusing the scope of inquiry down to those components that had direct relevance to the conduct of warfare and thus avoiding digressions into other, indirectly related and peripheral issues.

These theoretical generalizations retained a practical value only for the individual commander who had originally deduced them from his own critical analysis of history. If directly adopted by less experienced officers without the benefit of critical analysis these same generalizations would assume an abstract, objective quality difficult to translate into practice. However, in the conduct of tactics where the material aspect of warfare had a more direct and perceivable influence on the outcome of physical struggle the less sophisticated of these theoretical generalizations still retained practical value as prescriptive guidelines for action. Likewise, the evaluation of an army's material needs for conducting potential combat (otherwise known as strategic planning) could also benefit directly from these objective principles.

As responsibility for decisions grew with the corresponding growth in operational complexity that marked the difference between tactics and strategy it was imperative that this dependence on theoretical principles, or *routine,* give way to the practice of judgment. This progression away from the specific, technical knowledge of the subordinate officer to the synthesizing inspiration of the senior commander was to ultimately define the full concept of military art:

> War, in its highest forms, is not an infinite mass of minor events, analogous despite their diversities, which can be controlled with greater or lesser effectiveness depending on the methods applied. War consists rather of single, great decisive actions, each of which needs to be handled individually. War is not like a field of wheat, which, without regard to the individual stalk, may be mown more or less efficiently depending on the quality of the scythe; it is like a stand of mature trees in which the axe has to be used judiciously according to the characteristics and development of each individual trunk.[3]

It was very important, in light of the incidental nature of war, that the education of senior commanders and subordinate officers should instill in them the most critical-analytic ethic for evaluating past military experiences and theory of contemporary war. This theoretical tradition in conjunction with evolving institutional responsibilities of the high command and the general staff culminated in the concept of military doctrine that was to guide the performance of the nation's army in time of war. If constantly subjected to critical review it provided the professional soldier with a responsive tool, if left unchecked it degenerated into military formalism.

The purpose of this present work is to document the evolution of the military-theoretical tradition at the Nicholas Academy of the General Staff during a century fraught with tremendous technological innovation, wars in an increasingly tense international arena, and governmental reform in the face of growing social, economic, and political unrest throughout the Empire. Overshadowed by the Prussian reputation for excellence in the military profession during this era, and subsequently dismissed by Western military scholarship, the Nicholas Academy was nonetheless successful in promoting these educational ideals while contributing many new theories of its own. Although severely tested by the ravages of the First World War and the October Revolution this military-theoretical tradition was instrumental in the development of the Red Army and has retained its significance in Soviet military doctrine to this day.

NOTES

1. C. von Clausewitz, *On War,* edited by P. Paret and M. Howard (Princeton: Princeton University Press, 1984), 101.
2. For a more detailed discussion of these two military-philosophical traditions in the Napoleonic and post-Napoleonic eras see C. E. White's *The Enlightened Soldier: Scharnhorst and the Militarische Gesellschaft in Berlin, 1801-1805* (New York: Praeger, 1989) and *Jomini and His "Summary of the Art of War,"* edited by J. D. Hittle (Harrisburg, PA: Stackpole, 1965). A more general synopsis can be found in Martin van Creveld's *The Training of Officers. From Military Professionalism to Irrelevance* (New York: Free Press, 1990).
3. C. von Clausewitz, *On War.* 153.

RUSSIAN IMPERIAL MILITARY DOCTRINE AND EDUCATION, 1832–1914

1

Innovation: The Foundation of the Nicholas Academy of the General Staff under A. H. Jomini and Nicholas I, 1832–1847

Following the example of other European military establishments in the years after the Congress of Vienna, the general staff of the Russian imperial army gradually gained prestige and, on the model provided by the *Kriegsakademie* in Berlin and the *École Supérieur Militaire* at St. Cyr, the School for Column Leaders was established in 1819 by General Murav'ev to provide specialized training for the general staff. When Alexander I died and his son, Grand Duke Nicholas, took the throne in the spring on 1826, this school and the role it had played fell under the suspicion of the new tsar, not least for the reason that several general staff officers associated with it had challenged his legitimacy as tsar by participation in an attempted *coup d'état* the previous December.

Nicholas I's reasons for abolishing the School for Column Leaders and restructuring higher military education were not strictly political. In his former days as a lieutenant-general in command of the Guards 1st Division, he had often been angered by the insolence of those cadets who played an officious role in the conduct of annual maneuvers and had pledged to himself to only allow mature and experienced officers to serve on the general staff. On the occasion of his coronation, Nicholas I asked the great military historian and court military theorist General Antoine-Henri Jomini to draw up a new proposal for an appropriate educational institution to replace the School for Column Leaders.[1]

Although further proposals on the reform of Russian higher military education were interrupted by war with the Ottoman Empire in 1828, Jomini eventually convinced the tsar that another school founded on the special needs of the general staff would be less useful than an academy which admitted military engineers as well as artillery, infantry, and cavalry officers from the active army. His motives

Adjutant-General Baron Jomini

for sponsoring the foundation of this military-educational institution with a broader mandate were quite different from Nicholas I's original intentions. The academy was to serve a dual purpose: to provide the most intelligent officers in the army with the fundamental vocational skills required of general staff officers and junior commanders; and to provide a forum for the development of strategic theory. Jomini's past work had been concerned with the study of military history as the empirical source from which principles for a science of warfare could be developed, the importance of which he stressed in his correspondence with the tsar: "How could one deny the necessity of teaching a science which could have so much influence over the destiny of states and which often determines their grandeur or their decadence?"[2] This academic function of the academy was intended to research the geographical features of a zone of military operation, known as the theater of war, and to analyze its internal geographical features in conjunction with the material capabilities of the opposing army to prescribe the distribution of supply depots and the most advantageous movement of troops (otherwise known as the bases and lines of operation). The final goal in the development of this theory of strategic options was to simplify the determination of the most advantageous sector of the enemy's forces against which to concentrate one's forces.

Jomini was well aware of the enormous obstacle that the entrenched conservative service ethic prevalent throughout the tsar's administration and particularly strong within the military would present to the incorporation of professional priorities. In part he agreed with this traditional service ethic, which demanded total subordination to the innate talent of the supreme commander who embodied the spirit of the nation and thus represented a higher level of consciousness than those serving as his agents. Yet Jomini argued that this age-old ethic only comprehended half the truth about the nature of modern military operations:

> Doubtless, genius has a great share in success because it presides over the application of the established rules, and seizes all the shades of variety of which they are susceptible; but in no case will the man of genius act against these rules. An unskilled general may obtain a victory in opposition to the first principles of science; it happens so sometimes, for one of the two parties engaged must need gain it: but such an event proves nothing but a reciprocal incapacity, and total ignorance of tactics of the two chiefs. Such were the battles of the middle ages, in which the quality of the troops, and the bravery of the chiefs were the ordinary causes of success.[3]

Strict observance of formal etiquette and the inspiration of the tsar could no longer fulfill the professional obligations of the largest standing army in Europe under new conditions of warfare, nor should it deny the widest possible military-scientific training to its officers or risk preserving illiteracy within the ranks.

Without infringing upon the personal authority of the tsar, the academy was to attend to the training of an élite within the officer corps which was to serve the tsar with vocational precision and expert understanding of abstract military problems. Jomini stressed the importance of this in his correspondence with the observation that "a good general staff is as important to the army as a good ministry is to the state. Without it, one can have good regiments but not a good army," echoing almost verbatim his discussion of general staff officers' training in his previous work.[4] A general staff embued in a professional ethic of merit and trained in the theory of military strategy at a central military academy would provide the Russian army with the proper basis upon which to conduct modern military operations successfully.

Armed with the tsar's approval for a wider military-educational institution and the assurance that he would be appointed president of this future institution, Jomini drew up a provisional proposal which further defined his ideas. He suggested that the institution should take the form of an academy in the broadest sense of the word—a military-educational institution which would cater to the education of the entire officer corps of the imperial army—from staff captains to general officers.[5] This military academy would be under the general supervision of a director, immediately subordinate to the chief of the main staff, who would be responsible for chairing the academic council. A vice-director and two section chiefs would supervise the work of the professors and the day-to-day administration of the institution. The academy would seek to provide the training of staff officers using a progressive curriculum, the first step being preparatory course work laying down a firm foundation in mathematics for the study of topography and geodesy.

The main curriculum of the military academy would then begin with a year of introductory lectures on the following topics:

1. Contemporary tactics of the different services with the use of terrain in the execution of maneuvers.
2. The elements of strategy (or, as Jomini called them, grand tactics), consisting of his own theories concerning the complicated array of lines and bases of military operations.
3. The study of military history, comparing ancient modes of conducting warfare with contemporary ones.
4. The study of the military forces of various Western European nations in both their geographical and statistical characteristics (military geography and statistics).
5. The study of permanent and field fortifications, military administration problems, and lectures on the evils of false doctrine and incorrect theories.[6]

The second year of study was to be the practical side of the curriculum, allowing students to manipulate the previous year's lecture material through seminars and individual written research assignments. Jomini suggested that the main focus of these exercises should be to manipulate the information provided by the study of military history to form theoretical principles as well as to debate the significance of the military statistics and geographical distribution of potentially hostile nations, concluded by the "consideration of new inventions of foreign creation and their possible use in the Empire."[7] In addition to the function of the academy as an educational institution, Jomini again emphasized its role as a forum for the development of military-scientific research in the imperial army.[8] Although these ambitious proposals were generally accepted by the tsar and a special commission to be chaired by Jomini was eventually established, Jomini's proposals were subsequently modified to suit the conservative motives of the autocracy.[9]

The commission's work to draw up the details of the official decree establishing the military academy immediately showed the strains of disagreement between the chairman (Jomini) and the four members (Count Sukhtelen, Lieutenant-General Khatov, Major-General Schubert, and Adjutant-General Neidgardt). The main point of contention was the academy's role in the education of the Russian officer corps. The commission limited enrollment to candidates from the artillery and engineering schools and imposed draconian selection procedures which cut the enrollment of students down to all but a handful.[10] In consequence, the proposal for a year of preparatory course work preceding the regular curriculum was dropped, as was the proposal for using the academy as a forum for military-scientific research.

Jomini's recommendations for the appointment of the director and members of the faculty were also ignored. Instead of the appointment of either of the two foremost Russian military theorists, General Khatkov or Colonel Okunev, General Sukhozanet of the Guards Artillery was eventually appointed director. A severe disciplinarian, Sukhozanet had been the commanding officer of artillery at the Winter Palace on December 14, 1825, and had executed Nicholas I's order to fire upon the revolutionaries. During his twenty years as director of the academy, Sukhozanet was to maintain strict control over the administration and curriculum and was to seriously hamper the intellectual development of strategic theory.[11]

Otherwise, much of Jomini's model for the structure of the curriculum and administration of the academy was preserved in the official *Ustav* published on October 4, 1830. The director was to preside over the academic council in its review of curriculum and was to appoint lecturers, while the vice-director was to preside over the academic conferences which brought together the professors and their assistants to discuss course content and pedagogical methods as well as review the written exercises of the students. The curriculum was divided into a first-year "theoretical" section of lectures and a second-year "practical" section.[12]

The work of the commission toward implementing the provisions of the official *Ustav* was once again interrupted by events outside Russia's borders. In the autumn of 1830 Poland was disturbed by political unrest which had been reverberating throughout Europe. By early November a popular uprising in Warsaw caught the Russian garrison troops by surprise. The subsequent suppression of the Poles and the final assault on Warsaw in August 1831 was marked by incompetent leadership and extreme privation suffered by the troops. Only when the situation was safely in hand did General Chernyshev, both war minister and chief of the main staff, once again turn his attention to the severe deficiency in the number of staff officers in the Russian army. In late 1831 he hastily formed the academic council for the academy, sought the patronage of Grand Duke Mikhail Pavlovich, and on January 11, 1832, appointed General-Adjutant I. O. Sukhozanet as director of the academy.

Although Sukhozanet, as chief of artillery, had been seriously wounded during the recent war with Poland, he immediately began work to locate a building in which to house the academy and to seek the fulfilment of the commission's proposals for the proper administration of the academy. By the end of March 1832, preparations had progressed to the point where the academic council was prepared to devote its attention to the formulation of examination procedures, setting the calendar for the academic year, and assigning each faculty's lecture schedule for the academic week.[13] However, finding instructors for the academy proved to be another matter, for not one of the instructors from the other military schools responded to the call for academic personnel. Sukhozanet found that he had to more actively solicit instructors from other military academic institutions. He appointed Baron von Medem from the artillery school as professor of the faculty of tactics, who then took over the faculty of strategy, military history, and military literature; Colonel Viemarn from the Guards general staff as Medem's adjunct professor in tactics (a position which he inherited when Medem began to teach strategy); and Major P. A. Yazykov from the Institute of Engineering and Communications as professor of military geography and statistics.[14]

The fundamental work on developing the academic program was delayed until mid-September when the vice-director, Baron von Zeddeler, returned from a trip abroad to supervise the day-to-day administration of the academy and the formation of the academic curriculum. In temperament Zeddeler was the polar opposite of Sukhozanet. He had been educated in Austria and had begun his military career in the Austrian army where he had won distinction as a general staff officer. Upon joining the Russian army in 1813, he had served on the Russian general staff. Admired by his contemporaries, Zeddeler was well acquainted with military affairs and possessed a love for the study of military issues, which Jomini so highly valued.

In the first academic conference held on September 13 and October 18, Zeddeler called together the lecturers and their assistants to draw up the academic program and method of teaching military history and strategy for the first-year lectures. Addressing the skepticism widely held throughout the Russian imperial officer corps for the utility of strategic theory in the conduct of warfare, Zeddeler and the members of the academic conference proposed that the study of the history of military art, as opposed to the sequential study of events in the history of wars, should concern itself with

> the art of military considerations. . . . In order to attain thoroughness . . . one must study beside military considerations and action of the ancient military commanders the general form of wars in different eras, both ancient and middle ages, and of different peoples in different countries under different circumstances.[15]

ГЕНЕРАЛЪ - АДЪЮТАНТЪ
СУХОЗАНЕТЪ.

Adjutant-General Sukhozanet

This part of the curriculum, the history of military art, was very ambitious because it recognized a broad chronological and social empirical base as the appropriate source for the construction of military theory. In the discussion over formation of a theory of strategy, the academic conference proposed that such a theory could not be formed until the history of the origins and the gradual development of military art (strategy and tactics) had been accomplished, and that such an enterprise should be conducted following a military-theoretical-historical plan to "collate findings of practical aspects within military history with the results of research into the theoretical deduction of different strategic theories."[16]

This strong emphasis on the role of military history in the development of the curriculum was due to Zeddeler's influence. Zeddeler asserted that the history of military art was important

> because it serves as a practical introduction to military science, shows the advantages and insufficiencies of its laws in many different nations . . . and explains how these sciences attained, in the end, present perfection. Without such a fundamental knowledge of the history of military art it is impossible to understand military history, and without knowledge of that it will be impossible to conduct military education or have good commanders.[17]

Zeddeler thereby established the curricula of analytic military history as the theoretical core for all the disciplines to be taught at the academy.

Zeddeler's dominant influence over the creation of the curriculum soon caused antagonism between himself and Sukhozanet as well as with instructors. Zeddeler's feud with Sukhozanet concerned both the administration and curriculum; the director's attitude toward the academic function of the academy was unabashedly anti-intellectual with the maintenance of strict discipline as the first priority.[18] He had also infringed upon duties which, according to the official *Ustav* of the academy, were those of Zeddeler as vice-director, namely, supervision of the administration and academic program of the academy and appointment of academic personnel. This antagonism between the two was, for the most part, due to the imprecise wording of the *Ustav*, but in his attempt to regain authority over the affairs of the academy by imposing his military-historical-theoretical methodology on the development of the curriculum, Zeddeler managed to antagonize Yazykov, who attempted to exclude Zeddeler from the academic side of the academy's affairs in 1833 (the reason for antagonism was over the role of the academic conferences in the formation of curriculum). Zeddeler also alienated Medem, who did not like Zeddeler's concept of an independent history of military art.[19] Friction over control of the curriculum resulted when the higher academic council in late September 1832 decided to split the course on military history into two sections: the history of wars to be taught under the faculty of strategy, military history, and military literature; and Zeddeler's history of military art, reduced to instruction in service regulations under the faculty of tactics.[20]

ГЕНЕРАЛЪ - ЛЕЙТЕНАНТЪ
БАРОНЪ ЗЕДДЕЛЕРЪ.

Lieutenant-General Baron Zeddeler

Over the next two academic years Sukhozanet and Zeddeler squabbled over administrative and academic issues until late September 1834, when the final collision occurred, precipitating the dismissal of Zeddeler from the academy and the Russian imperial army for insubordination.[21] In such a manner the tone was set for the development of Russian military theory and knowledge in the twenty years up to the commencement of the Crimean War. As a historian of the academy was to later note, life as a student in those early days was extremely spartan, exemplified by the difficulties of paying attention during afternoon lectures—"especially in the depths of winter when the days were short, and in those days, no one considered lighting the class rooms."[22]

The first great military theorist to contribute to modern Russian military thought during the early years of the Nicholas Academy was Baron von Medem. In assuming the lectureship of strategy, military history, and military literature in 1833/34, Medem took on the enormous task of providing instruction in all of these fields within the confines of three lectures (six hours of instruction per week). His lectures, later to be compiled and published as the *Review of the Known Laws and Systems of Strategy* (*Obozrenie izvestneishchikh pravil i sistem strategii*) (1836), introduced the subject of strategy with a short, general history outlining the increasingly complex nature of warfare caused by the introduction of new means of combat—from the small armies of the ancient Greeks conducting single decisive battles to eighteenth-century European armies armed with firearms and preoccupied with the protection of their supply lines and depots.[23]

For Medem such a historical review of the changing forms of conducting warfare served to indicate a general trend over the ages: "Strategy as an art has always existed, strategy as a science gained significant importance in military operations when they became more complex and difficult." This fact posed the challenge of creating a systematic code of ideas concerned with the growing complexity of military affairs.[24] Having introduced his students to this review of strategic systems of thought, Medem launched into the second part of the course, that of a descriptive review of strategic theory as expressed in great campaigns, or what was known at that time as military history. In this section a general review of the military campaigns in the classical era was covered, while significant time was devoted to the more recent campaigns of the Seven Years War and particularly Suvorov's campaigns. An attempt was made to approach the subject matter from many perspectives in order to discourage students from falling into a routine or narrowmindedness.[25] However, to place too much emphasis on the distillation of past military experience in order to deduce mechanistic general laws regulating military phenomena would have had, in Medem's opinion, a harmful effect on the student's proper understanding of future military operations. At the end of the course Medem conducted a short review of military literature with the purpose of stimulating the graduating student's further study of military knowledge.[26]

The most significant part of Medem's course was in the first part of his lectures, in which he presented an analysis of other strategic theories showing how a priori orientations to the study of the history of military art on the part of Lloyd, Bulow, and Jomini had led to mistakes. Medem showed how Jomini drew from the military experiences of Frederick the Great and Napoleon Bonaparte to garner two sets of laws: Frederick's system of using interior lines of moving troops which demanded the division of troops into separate units, and Napoleon's concentration of troops for a single, decisive engagement.[27] Jomini had tried to resolve this disharmony by claiming that geographical features may require several different uses of operational lines of movement, thus providing the theoretical basis for his "geometry of military art." For Medem, such a single-determinant theory distorted the significance of geographical factors. Medem's conclusion from this analytic study of Jomini's strategic theory as well as other single-determinant theories was rather disparaging: "constant, indisputable laws for successful military operations do not exist."[28]

Medem's pessimistic attitude toward an applicable strategic theory, later noted by War Minister D. A. Miliutin (at one time a student of Medem), was founded on his materialist concept of the development of military art.[29] In his opinion the developing technology of warfare had a tendency to radically change the configuration of the battlefield: "The invention of some new weapon, using new means up till now unknown, can suddenly change everything, and it is impossible to certify that strategy does not experience such change when it undergoes the introduction of a new weapon," thus making any attempt to form eternal and unchanging laws of warfare thoroughly pointless.[30]

More pointedly, Medem's fundamental disagreement with the popular historical school, which sought to establish a mechanistic system of principles to explain social activity, lay in the various possible definitions of the word "science." If the word simply meant an epistemology consisting of the considerations taken into account for the successful conduct of military operations, an abstract theory ever-changing in accordance with available means and circumstances, then such a concept "shows that we are coming closer to understanding the essence of science."[31] For Medem the only utility of teaching the theory of strategy and conducting a historical review of military theory was to allow the young officer corps to become "acquainted with the characteristics of all strategic circumstances and means, estimate their mutual influence and to research the importance of each separate element in regard to military operations."[32] Such an exercise was not to provide a strict guide to the proper conduct of military operations—as Lloyd, Jomini, and the historical school had attempted—but only a stimulus for the commander's creative abilities in time of war. Medem's argument thus echoed von Clausewitz's discussion on the training of the commander.[33]

In *Tactics* (*Taktika*), published two years later, Medem defined tactics as the "art of forming troops and their use in the fulfillment of strategic considerations," which provided the continuation of a theme he established in his previous work: the component elements in the theory of strategy.[34] The highest component of strategy was the commander's awareness of the material and moral strength of his troops and his capability as a commander to successfully manipulate them to meet the circumstances of the battlefield. Introductory tactics was to "examine those material factors of armament, organization, movement, and action of each type of troops," outside the context of terrain, while applied tactics examined those factors within that context.[35] With the course of the lecture on tactics set, Medem then examined the basic concepts common to all types of troops: the meaning of offensive and defensive goals; the role of discipline and the importance of various factors making up spiritual strength; the division of weapons into firearms, cold steel, and preventative (defensive) with the firearm component further divided into handguns and artillery; the division of troops into regular and irregular, with the infantry, cavalry, and artillery services; and the distribution and movement of all three services in battle.[36] In such a manner a distinct regard for the analysis of military art, or the military "genius" of the commander, was to take its place in discussion of the appropriate composition of military science.

Medem's lectures were to have a profound influence on the future development of strategic and tactical theory and the practice of military history at the Nicholas Academy. His refutation of the purely deductive orientation to the development of military theory was to haunt Russian military thought up to the October Revolution and beyond, and his fundamental regard for the role developing technology had to play in the development of theory was well ahead of its time.

Although a course on military geography and military statistics was specifically called for in the official *Ustav* of the academy, no model for such a course existed in either Russia or Europe. The task of creating such a course fell to Major P. A. Yazykov, formerly of the Institute of Engineering and Communications. Yazykov initially turned to works by Prussian and Austrian military geographers, such as Lieutenant-Colonel Rudtorfer[37] and others, but they proved to be of limited value. The works of Jomini, especially his *Treatise upon Grand Military Operations* (*Traité des grandes operations militaires*), expounding the concept of bases and lines of operations, as well as the works of the distinguished veteran of the Patriotic War, Colonel Okunev, provided a more analytical model on which Yazykov based his course outline.[38] Although the course on military geography and statistics was originally allotted four lectures (eight hours per week), by the first meeting of the academic conference the lack of any theoretical work devoted specifically to military geography and statistics, the very small base of literature, and the lack of appropriate teaching aids other than topographical maps necessitated cutting the course down to only two lectures on military geography per week and an occasional lecture on military statistics.[39]

In the first year of this course Yazykov found the task of teaching and simultaneously creating such a course too much for one person and applied to the academic council for an assistant. After the first assistant fell into disfavor with the director, Lieutenant-Colonel Stephan was hired in 1834 to help Yazykov teach the course. The content of the course was to consist of two parts. The first was what Yazykov called the descriptive or purely empirical part of the course and was concerned with geographical data, while the second part was the analytical discussion of military-historical experience, which was to elucidate the relation of the empirical data to defined goals of military thought. Both aspects formed the military-geographical review of a particular nation or theater of war. Yazykov provided such a review of the Russian Empire, Prussia, and Persia to the first-year students, while Stephan did the review of four theaters of war in the Austro-Hungarian Empire, European Turkey, and Sweden for the second-year students.[40]

In the introduction to his *Attempt at a Military Geography* (*Opyt teorii voennoi geografii*), published in 1838, Yazykov defined the need to perform an inquiry, stating that since no one had adequately defined this important element of strategy there was no reason why a Russian could not attempt it.[41] Yazykov had developed some ideas about the character and role of military geography, as can be seen in his typology of military-geographical research into the descriptive and the analytical when the course was first introduced. It was not until the publication of Jomini's very influential work, *Summary of the Art of War* (*Précis de l'Art de la Guerre*) in 1837 at the request of Nicholas I (and which Yazykov translated into Russian) which defined in more accurate terms his concept of strategy, that Yazykov was stimulated to develop his own theories about military geography and its place in military thought. As far as Yazykov was concerned, Jomini's *Summary* was the most sophisticated attempt at a theory of strategy in contemporary military literature. Medem's *Review* had done a good job of pointing out the weaknesses of the contemporary theories, even those of Jomini's earlier works, but Jomini's subsequent reevaluation of his theory in the *Summary* had surmounted some of those obstacles through a thorough analytical review of military-historical experience. Others, in contrast, had relied exclusively on a purely descriptive analysis of either geography or the history of wars.

Yazykov was greatly influenced by Jomini's original assertion made in his *Treatise* and repeated in the *Summary,* that "there exists a small number of fundamental principles of war" revealed by a study of the historical works of great commanders such as Alexander of Macedonia, Frederick the Great, and Napoleon Bonaparte.[42] Yazykov believed that Jomini's theories were attuned to practical application rather than convoluted abstractions. The key to understanding strategy was to realize that no single factor was involved; it was a complex amalgam of different elements, some of which belonged to the natural sciences and some to the social sciences. To Yazykov the brilliant example of this dual nature within strategy was Jomini's category of military policy under which were the studies of military geography and statistics involved simultaneously with diplomatic, moral, and topographical factors.[43] This dual nature of Jomini's *Summary* was unlike the predominantly a priori nature of his earlier work, especially the *Treatise.* In writing the *Summary,* Jomini remained sensitive to those criticisms of his earlier work, especially those of von Clausewitz, for he moderated his demands on the strictly normative nature of the principles of war:

The maxims of application which are revived from these principles are also small in number, and if they are found sometimes modified according to circumstances, they can nevertheless serve in general as a compass to the chief of an army to guide him in the task, always difficult and complicated, of conducting grand operations in the midst of the noise and tumult of combats.[44]

The mechanistic universal laws of the natural sciences might not be applicable to social phenomena, but the study of military history would point out the recurring patterns from the specific details of past military experience, those issues which were to remain eternally salient for the consideration of the commander. For Jomini as well as Yazykov the role of analytical history was central. Its goal was to pare out the dead wood of circumstantial phemonenality of past military events in order to reveal eternally recurring patterns or truths. Those issues which retained a comparative quality of stability over others and which were readily apparent were geographical truths, and it was up to the genius of the commander to constantly review the interaction of all other circumstances with these eternal and perpetual truths of physical geography. Of course, Yazykov was aware of this analytical historic-geographical orientation. Heartened by Jomini's hypothesis that the analysis of physical configurations of the landscape in relation to consideration of strategic movement of troops could possibly be a central element within the dual nature of strategy, Yazykov sought his own theory of strategy in a work published a year after Jomini's *Summary.*

Yazykov began with the fundamental philosophical issue which he thought was causing the confusion in the creation of a theory of military geography, and, by implication, strategy. Yazykov identified two types of knowledge: the first was purely empirical (descriptive methodology); the other was far more demanding and elusive (the analytical method concerned with the interaction of means and circumstances with the empirical physical configuration of the land).[45] To illustrate this important dual aspect of the military sciences, Yazykov embarked on a critique of German essays on military geography, especially those of Rudtorfer, relegating them to the confines of purely descriptive general geography and statistics.[46] This was not to deny the importance of these empirical sciences (although Yazykov would have objected to calling them sciences) but merely to point out that they did not satisfy the needs of a theory for military geography or strategy: "Above all topography includes in itself many changing givens, and for this reason of a multiplicity of detail which would be impossible to know at a single time, topographical knowledge is only good for making maps."[47] Although Yazykov conceded that Rudtorfer noted the applicability of military history to the study of military geography, he pointed out that Rudtorfer's use of military history in his study of military geography was merely to provide examples of terrain as the most important determinant in military affairs. It was at this point in his critique of Rudtorfer and other contemporary European military theorists that Yazykov began to outline his own approach to the formation of theory:

There is not any doubt that the lamp of History illuminates the concept of Military Art. Strategy in particular, not being lit up by History would remain forever a labyrinth as many attempts to master the theory of Strategy have shown. But to describe geographical subjects and topographical ones as a simple review of those places where important events occurred which were formerly always reviewed by History one would not illuminate any strategic concepts.[48]

He dismissed Rudtorfer's work with a final barrage of criticism, pointing out the fact that many geographical features of a country do actually change. Roads, towns, forts, and borders can change quite quickly; this fact nullified Rudtorfer's argument that the physical configuration of the landscape was capable of providing eternal principles upon which to build theory.[49]

The place to start, then, in compiling a proper theory of military geography was to discern which geographical givens hold true for every historical epoch (the natural elements) and those which change (the artful elements).[50] Yazykov made an interesting connection between artful geographical elements and the social aspect of the nation: "The development of social activity not only gives birth to artful geographical subjects but even changes the external appearance of 'natural' geographical subjects and its means concerning the movement of armed forces." Unfortunately, he did not elaborate what he meant exactly by this statement.[51] Perhaps he was hinting at an appreciation for the significance of developing technology on the geographical distribution and coordination of a country's military forces, a notion which would have not been new to him as it had been the underlying thesis of Medem.

For Yazykov the military sciences, including military geography, were speculative sciences. As such they could not be expected to attain the simplicity or clarity of mathematical sciences. But if they were thoroughly and continuously researched and reevaluated, one could expect a high degree of certainty from the patterns (or principles) garnered from historical analysis, and on this basis could therefore claim that these speculative sciences were characterized by eternal and unchanging circumstances. This was consistent with his notion that the speculative sciences were bound to historical periods. The works of Lloyd, Jomini, and Montholon had correctly exercised this mélange of methodologies according to the historical circumstances of their age, but had subsequently become outdated and therefore obsolete. Having the best of both worlds in the dual nature of the so-called speculative truths meant, for Yazykov, a concession to the demand for practicality while at the same time preserving the significance of abstract theory. This was his central and everlasting contribution to imperial Russian military thought: "[from historical fact] it is apparent that truths do not change, only the givens. Consequently, givens do not constitute the character of truths which must be eternal. Therefore, it is necessary to define Military Geography as consisting of the unchanging character of speculative truths."[52] An implication of this observation was that the research methodology suggested by the study of military geography would also be eternal and unconditional for it would not be affected by specific circumstances.[53]

Elaborating on Jomini's description of military geography and statistics, Yazykov proposed that the eternal research method would regard the following: (1) geographical situation of the entire state or country, examining the totality of its geographical composition in comparison with other states; (2) form of geographical composition of the state; and (3) possible theaters of military operations, based on the physical characteristics of the country. One would consequently strive to explain how each of these geographical features influences a state's strategy.[54]

The advantage of a speculative science was that it recognized the differentiation from, and, at the same time, the integration of abstract theory with a theory of practical action, which Yazykov defined as the theory of art. He understood the importance of the genius of the commander in the manipulation of events in the heat of battle and, like Jomini, was careful not to devalue its important role in consideration of military science and military art. After all, the goal of military science was to complement this genius of the commander in the growing complexity of modern warfare and to provide efficient utilization of force to gain the edge on victory: "That person who acts instinctively can arrive with confidence at the goal found in circumstances, especially when he meets a case for which to execute goals he must reveal new means. . . . In order to fully know art one must comprehend the causes: [he is to] do that and nothing else."[55] For the student—and by this term Yazykov meant not only the young officers who made up the general staff but also the commander of the army—the best way to exercise one's talents in the art of warfare was to draw upon two sources: the observation of practice only possible in time of war, and the study of essays on military art written by great leaders. From studying these sources the student of warfare would come to appreciate the fact that the successful conduct of military art on the battlefield requires knowledge of the possible means at one's disposal, the method of their application on the battlefield, and the various methods of scientific speculation in the application of this military art.

The totality of all these concepts would provide the future commander with the full theory of military art, and when combined with field experience would provide the commander with that virtue Yazykov valued the most: decisiveness. Unfortunately, Yazykov's work did not enjoy critical acclaim at the time of its publication either within the Nicholas Academy or in the military press. The relative disfavor of Yazykov's theory on military geography and statistics found expression in the fact that when the editorial board for the *Military Encyclopedic Lexicon* was preparing an article for the entry on military geography and statistics in 1837, Yazykov did not contribute.[56]

A catalyst for change in the curriculum at the end of the decade was adjunct-professor N. S. Golitsyn's proposal to restructure the curriculum for the instruction of strategy and military history. Golitsyn's three years as assistant to Medem and colecturer had impressed upon him the opinion that "the more one studies history, the better one understands strategy," and that perhaps both could be taught more effectively if combined in a course which showed the close links between abstract strategic principles and their practical manifestation in historical events. In early 1838 he drew up a detailed proposal suggesting that instead of teaching a course which was completely abstract in content a new course should be inaugurated which would conduct a thorough examination of practical historical experience focusing on the state apparatus, foreign relations, and the arrangement and means of "utilizing the military forces and means of the government."[57] Golitsyn informally sought the advice of his mentor, Medem, on the practicality of this proposed change in curriculum. Although Medem approved of the need to combine military history with the study of strategy, he was well aware of the enormous parameters of this research and could see that the formation of such a course would demand an enormous amount of research which would take a long time to complete. Medem tried to dissuade Golitsyn from undertaking this task, but Golitsyn did not follow his mentor's advice; he submitted the proposal to Sukhozanet who, in turn, ignored the standard procedures for submission of proposed changes in curriculum to the academic council by submitting it directly to the war minister for approval.[58] Sukhozanet's reasons for lending so much support to Golitsyn's proposal may have been motivated by his desire to make the curriculum conform to more practical goals. The wish to undermine Medem's influence in the affairs of the academy, however, was probably stronger in light of his forthcoming retirement from the faculty of strategy and membership on the academic council. When Medem retired in December 1838, Golitsyn was appointed as professor of strategy and immediately created a collective editorial committee from the more talented of the academy's recent graduates to begin compiling material for the new course. Many of these editorial assistants, in particular M. I. Bogdanovich, D. A. Miliutin, and F. I. Goremykin, also contributed to the new *Military Encyclopedic Lexicon* (*Voennyi entsiklopedicheskyi Leksikon*) which was being compiled by former vice-director Zeddeler.

Although not exceedingly popular with this new generation of instructors, Yazykov was now the most senior lecturer at the academy; he published his essay on military geography which suggested a methodology for a unified theory of strategy inspired by Jomini's *Summary.* Yazykov gave lectures on strategy, published in 1841 as *An Attempt at a Theory of Strategy* (*Opyt teorii strategii*). In Yazykov's mind, the only reason Medem may have doubted the existence of an applicable theory of strategy was the admittedly primitive state of research.[59] In his new work he attempted to illuminate the darkness by offering a proposal on the direction future research should take, not unlike the academic conference's proposal or Medem's suggestions in his *Review:* "A theory of strategy can only exist when three conditions have been met: a). research is conducted on each separate element, b). the interrelationship between those elements is classified, c). when the total influence of all elements on practical affairs is explained."[60] In any event, Yazykov's first priority was to typologize the theory of military strategy into its theoretical and applied aspects, focusing attention on the development of the theoretical. Within the theoretical a further delineation was made by distinguishing between physical and psychological elements, much the same way Medem had done in his work on tactics. To place the consideration of psychological elements in equal light with that of physical elements involved in military phenomena was original for its time and set the stage for growing concern with the psychological element in both the troops and the commanders.

In Yazykov's opinion the most significant of the psychological elements influencing strategic military operations was that of politics—by which he meant foreign politics. To illustrate his point, Yazykov incorporated the historical example of Frederick the Great's military preemptive campaigns against Austro-Hungary before it could arrange a coalition war against Prussia. From this he drew the following conclusions: "political circumstances can define the appropriate moment for initiating military operations" and "political circumstance can influence the main direction of their use."[61] This discussion included no concept of domestic politics determining the nature of military operations nor did it depart from Jomini's limited concept that politics ceased to affect military operations after they had begun. However, it was the first of Russian military theories to place politics so high in the pantheon of military thought. Arguably, the second psychological element—relations of the government or the leader to the troops determining their discipline and spiritual strength—could have been a primitive concept of domestic politics having influence over military performance. Yazykov did not add much to Medem's quite thorough examination of this element except to note that it was not to be confused with the previous concept of diplomatic politics.[62]

As there were two levels of analysis involved in discussing diplomatic political matters and the moral bonds between the soldier and the state, so there were two types of strategy directly corresponding to them: "It is necessary to differentiate the main political goals of wars born out of political circumstance from the immediate strategic goals arising from the conduct of military events and subordinate to the main political goals."[63] Yazykov asserted that the political goals of the war could be either offensive or defensive in nature, much as Medem had argued in his work on tactics. But Yazykov went further by stating that the organization of the armed forces could also determine whether the war for the respective country would be conducted along offensive or defensive lines. He even argued that if the war were a large one, as was the case in the Patriotic War of 1812, the discipline and spiritual strength of the entire nation would determine its offensive or defensive conduct.

In his introduction to the work on strategy, Yazykov acknowledged the debt owed to Medem and von Clausewitz for proposing that the study of strategy should be focused primarily on understanding its elements and the nature of the relationship between abstract theory and practical action, especially the denial of eternal and indisputable laws governing the interaction of these elements. However, Yazykov was to praise Jomini and Okunev for revealing "such new truths which could have immediate application to practical action," some of them later described as the "influence of a theater of war's physical features on the conduct of military operations."[64]

Such was the nature of the contradiction between the two most fundamental parts forming Yazykov's concept of strategy: moral factors which cannot be described in a mechanistic manner, and physical factors which appear to be far more stable and conducive to description by more concrete interrelationships. This internal contradiction did not go unnoticed by Yazykov. The first classification which he made separated the two for individual analysis. Although Yazykov's treatment of the moral elements was innovative for its time and received more attention than it had in previous Russian military-theoretical texts, Yazykov was far more familiar with the physical ones. In what could be interpreted as the military geography/statistics component of his discussion on strategy, Yazykov gave a list of various elements which would fall into the category of physical elements to be examined: "1). Geographical features, 2). Strategic features, 3). System of recruiting the army and its armament, and 4). Numerical composition of the army with regards to the distribution of the armed forces in the Theater of war," each element of which received subsequent detailed discussion.[65]

However, physical and moral elements had been differentiated only for the purpose of clarifying the proposed direction of intended research. Yazykov was very concerned with the way they might interact, and devoted a section of his work to discussion of how the organization of the army, for instance, might influence the political form of the military operations (its offensive and defensive nature). He argued that all types of military operation were not feasible without prepared reserves. Geographical distribution of strategic and tactical supply depots were greatly influenced by perceived political objectives, giving a highly predetermined structural quality to the type of military operations the respective armed forces would be able to perform.[66]

In many ways Yazykov's contribution to the early development of military theory at the Nicholas Academy was diametrically opposed to the teaching of Medem; namely, his aspiration to develop a unified, positivist theory of military art based substantially on Jomini's works. For this reason Yazykov was to prove to be a highly influential figure in the development of military theory at the academy. He introduced the subject of military geography and statistics which was in turn later responsible for inspiring the formulation of strategic planning and the theoretical base for the territorial military administration system of military districts instituted in the 1860s.

The retirement of the academy's most venerated instructor in conjunction with the appointment of new lecturers precipitated a wave of conservative reaction in the military press against the teaching of military history and theory. Conservative groups advocated a strictly historicist orientation to the nature of history, insisting that the personal qualities of a tsar determined the overall nature of historical events in conjunction with geographical factors in the development of military art, not the socioeconomic characteristics of a nation or the development of technology. Therefore, all efforts to use history as a base for deducing a theory of strategy were to be dismissed.

For the most part the new instructors were able to ward off these attacks in the press by their combined effort in the publication of the *Military Encyclopedic Lexicon* which defended the role of history as the foundation of military theory and practice, an indication that this new standard of competence had been well assimilated by Medem's and Zeddeler's protégés.[67] Several of the instructors and graduates from the academy also contributed independent articles to the press and thus bolstered the "liberal" position of the academy. The most significant of these was an article by D. A. Miliutin, soon to become a major figure in Russian military affairs, on the enigmatic eighteenth-century commander V. A. Suvorov. Hailed in subsequent Russian and Soviet military historiography as the first to use historical analysis in the study of military art and as the inspiration for a flood of literature on this great Russian commander, Miliutin had trouble with the censors over its publication and his next contribution on a similar theme never saw the light of day.[68] Impressed by Medem's and Yazykov's opinion that psychological elements as well as material ones played an important role in the conduct of warfare, Miliutin referred to the theoretical works and correspondence of Suvorov as well as various accounts on the conduct of his military campaigns to show that his success in battle was due to a well-defined sense of judgment and respect for his men unhindered by royal prerogative.[69]

After completing a tour of Europe in 1846, Miliutin took over the teaching of military geography and statistics from Yazykov, his former teacher. Inspired by the many impressions he gained from his travels and his study of the Prussian army, Miliutin began to reevaluate the course material in an effort to forge from it a more widely based, objective, and practical course. He was appalled by the lack of any coherent structure in the course left to him by Yazykov, and in reworking the lectures on the military geography of Prussia Miliutin was to reveal his fundamental misgivings about the course's content:

> The more I gathered material for such a work, the more I read and pondered, the more I became convinced that it was unthinkable to create a special military "science" from purely geographical matter. The available and more or less successful attempts at a strategic analysis of the separate theaters of war also were unable to create such a science, and seemed to me rather like teaching aids only, empty of real basis as long as these analyses did not make use of all existing factors which comprised military means and strengths of warring states. From this I came to the conclusion that strategic analyses of . . . territories which might become theaters of war comprised only one of the many aspects of a general study of the military capacity of nations. Thus only a well-rounded study could establish the subject and [attain] the goal of scholarly teaching. In such a case it would be already not military *Geography,* but a special offshoot of Statistics, which might be termed "Military Statistics."[70]

In his *First Attempt at Military Statistics* (*Pervye opyty voennoi statistiki*), which was published in 1847 on the basis of his lectures, Miliutin expressed his opinion that military geography/statistics as a study of the phenomena of war and most aspects of military affairs was to gain its epistemological base from the social sciences. The general discipline of the social sciences had two underlying philosophical assumptions. The first was what Miliutin called "theoretical" and which in his opinion was inappropriate because it assumed to form laws binding the development of societies to accomplish a priori goals. The second assumption was more scientific and thus more acceptable. It represented a "practical" approach, namely, an attempt to understand the processes of complex social phenomena and to provide direction in improving or enhancing certain aspects of that growth.[71]

With such a purpose in mind Miliutin then made a distinction which was highly original and quite different from the orientation of both Medem and Yazykov. Recognizing the utility of a diachronic methodology particularly well suited to study the development of one phenomenon over the course of time, Miliutin argued that there also existed an equally viable methodology in the social sciences that would examine a group of phenomena at a given moment in time, namely, the statistical method. It was this new method of military-scientific research which was to be his center of attention—the study of military statistics.

In the introduction to the topic, Miliutin gave his definition of military statistics: "the research in a given moment of a government's strengths and means in a military context." Miliutin then elaborated on what he meant by a government's strength:

> Not only [concerned] about troops nor even about the armed section of the nation, but with the totality of a government's means and abilities necessary for the conduct of war, offensive or defensive. In order to truly know the might of a nation it is insufficient to examine only one geographic given and the number of its troops. [One] must research the sources of military means of that nation, its military institutions, and terrain features of the country in all its weak and strong points.[72]

In order to approach such a large field of study, Miliutin established a typology. The first category was the general study of the political destiny of the state and its development, including a description of the resources and natural characteristics of the territory of a nation, demographic characteristics, the structure of government, and financial issues. The next category focused on the composition and organization of the armed forces, both ground and naval. Miliutin provided some examples of the questions to be addressed in relation to this theme, among them the comparative advantages and disadvantages inherent in the system, how fast the army could mobilize to full battle-readiness, whether the war was to be offensive or defensive, and a discussion of the theory of military art and the main ideas forming military actions.[73]

However, the final and most complex part of the typology was the research into a government's strategic plan for military operations in a particular theater of war. The concept of the theater of war was very important to Miliutin in his systematic attempt to understand and manipulate the course of military affairs. This theater of war was to be formed by a consideration for homogeneous geographical features overlapping domestic and foreign soil followed by an examination of the geographical features within that region. With such a geomilitary regionalization, the full collation of military statistics for that given area could be used in formulating a war plan of sorts and for addressing major questions of essential military infrastructure, such as where to put railroad networks and where to establish operational lines of communication and bases. This war plan was to take in the full spectrum of the military sciences and military art, and was not to rest on the example of historical events alone. Miliutin's criticism of Yazykov's tendency to mix the study of a priori history with the study of military geography was yet another clue to his concept of the military-scientific process: "The historical example is helpful to strategic research of theaters of war only in conjunction with a military-statistical review of tangible and intangible resources."[74] In this respect Miliutin made a brilliant analogy: "Military statistics is like the last page of military history; it is the history of the present, thus providing that essential link between theory and practice."[75]

To support these new goals of military geography, Miliutin initiated a military-geographical-statistical review of forty-three provinces (*gubernii*) and districts (*oblasti*) in the Russian Empire which was to provide material for future strategic planning.[76] These theoretical and practical efforts, hand in hand with his efforts to promote geography as a vehicle for the study of the Russian Empire's larger social issues (such as the emancipation of the serfs) in the Imperial Geographical Society, were to be the first systematic steps toward strategic planning and later hailed as the first truly scientific work in Russian military literature.[77]

A contemporary of Miliutin was M. I. Bogdanovich who, as professor of strategy, military history, and military literature, was to also play a significant role in promoting the liberal school of military thought well into the 1860s. As a cadet at the academy in the first years of its program, Bogdanovich had attended Medem's lectures on the development of strategic theory and the practice of military history alongside contemporaries such as Goremykin, Miliutin, and other figures who were playing a significant part in the development of Russian military thought.[78] In 1838, only two years after he had graduated from the academy, Bogdanovich was chosen by Golitsyn to be his main assistant in the collective effort to compile a course on a unified strategic theory, finding himself in charge of the project only a year later. Bogdanovich also participated in the collective editorship of Zeddeler's *Military Encyclopedic Lexicon* and proved to be an effective supporter of the continued instruction of military theory and history.

Like many of his peers who had studied strategy and military history under Medem in the mid-1830s, Bogdanovich was greatly influenced by Medem's ideas. In his first major work, *Essays on Strategy: Rules of Conducting War* (*Zapiski strategii. Pravila vedeniia voiny, izvlechennye iz sochinenii Napoleona, ertsgertsoga Karla, generala Zhomini i drugikh voennykh pisatelei*), which was published in 1847, he used the same review technique in describing the development of strategic theory which Medem had used ten years earlier. Bogdanovich also shared with Medem, Yazykov, and many other military theoreticians of that time the notion that strategy as an art had existed since the dawn of time but that it had only recently taken on the demands of a science due to the growing complexity of modern military operations.[79]

In interpreting this fundamental statement, many of Medem's students, including Miliutin and Goremykin, had accepted the dual nature of strategy and had even adopted Medem's bias toward the implications that emerging technology had in the conduct of military operations. Bogdanovich's work, however, reflected a strong influence from von Clausewitz's concern for the personal merits and judgment of the commander, an influence which was later to be labeled as subjective.

Instead of accepting the more potentially relative dualistic concept of strategy which inherently shunned eternal and unconditional laws, Bogdanovich believed that "the main basis of military affairs for all time remains unchanged." He accepted the critical role a commander's judgment played in the outcome of battles and eventually in the evolution of military theory on strategy.[80] For him, the main focus of military-historical analysis was to use official documentation to reveal the eternal and unconditional principles of military art through the study of the genius, or personality (*lichnost'*) of the commander, especially that of military monarchs: "Peter I, Charles II, Augustus [Frederick] II were heros (*bogatyrs*) who decided the fate of their nations through the reflection of their personal attributes in the ability to conduct war."[81] In pursuit of this subjective orientation to the progress of history, Bogdanovich was to write many histories of famous commanders, such as Napoleon Bonaparte, Suvorov, and Rumiantsev, which were to occupy an important place in the development of Russian imperial military historiography throughout the rest of the nineteenth century.[82]

Bogdanovich's *Essays on Strategy,* which followed the lecture outlines for his course on the theory of strategy and military history during the 1840s, offered a review of the main contemporary theories of strategy as Medem had done, but with the single and very significant difference that he commented on von Clausewitz's *On War,* thus introducing the last and most significant work by von Clausewitz to the Russian military community.[83] Bogdanovich deplored the unfinished quality and convoluted argument of *On War:* "to propose such unclear laws for guidance in the art of conducting war is similar to if one were to pose an undefined mathematical question; one would get many different answers."[84] However, the work was invaluable in providing a deep understanding of historical facts in combination with practical experience to unify the development of theory.[85] That he was very taken with this approach was to be seen in the statement he made in his article on strategy in the *Military Encyclopedic Lexicon:*

In our times there is no doubt that the *single* way to form a science of strategy is to be found in the fundamental study of military history, and that the theory of strategy must be limited to the examination of different means in the hands of the commander and the definition of the interrelationships and total influence of all strategic means and elements in military activities.[86]

The outcome of Bogdanovich's own attempt at a unified theory was a collection of principles which were to represent an implicit military doctrine. The first principle was one which was originally introduced into Russian military-theoretical literature by Jomini's *Summary* (which, in turn, had been influenced by von Clausewitz's work): the important role political guidelines had to offer in the planning of military operations,[87] considerations which were to specify whether an offensive or defensive military posture was to be practiced. However, Bogdanovich did realize that they were not necessarily mutually exclusive as an army may have to resort to a temporary defensive posture in order to muster resources to carry on the offensive.

Bogdanovich felt that the offensive was the superior posture and his rationale for such an opinion was that the defeat of the enemy's army or the acceptance of one's own will was the goal of military operations, a point of view which he shared with von Clausewitz.[88] Decisive armed engagements were to be the main goal for the commander as far as Bogdanovich was concerned, demonstrated by Napoleon's many great successes gained by concentrating his forces against the most important sector of the enemy's forces.[89] In determining where this important point was and the appropriate way to bring one's forces together, Bogdanovich borrowed Jomini's concept of operational lines, approving of the work Yazykov and Miliutin had done on choosing the correct line of operations using selected geographical and statistical criteria.[90] This implicit reference to the role of military geography and statistics was the closest Bogdanovich came to acknowledging the important influence material factors could have on the conduct of warfare. The predominance of eternal subjective elements in the theory of strategy, namely, the promotion of the commander's personal qualities and capabilities, remained the central focus of his work.

Other proponents of the liberal orientation who were to contribute their opinion to the development of military thought at the academy were Captain N. D. Neelov, assistant lecturer under Bogdanovich at the faculty of strategy, military history, and military literature, and Colonel F. Goremykin at the faculty of tactics. Neelov contributed many essays to the literature on the theory of strategy, culminating with the publication of *A Study on the Composition of Contemporary Strategy* (*Ocherk sovremennogo sostoianiia strategii*) between 1846 and 1849. Believing strategy to be a systematic code of theory to aid the art of the commander, he supported the materialist orientation of previous instructors, emphasized in the statement that "the genius of the commander having influence over circumstance, will in turn be influenced by circumstance."[91] He also shared the opinion that the only tool to assist in the study of objective military science was history, and because the conformity of strategy as well as tactics was based on ever-changing means of warfare one could not gain utility from the study of the ancient Greeks. Like his superior, Bogdanovich, Neelov felt that the study of Napoleon I's military campaigns had much to offer the student, not so much because of Napoleon's personal qualities as a commander but because of the superior quality and quantity of troops revolutionary France could field over those of the feudal-absolutist states.[92] In Neelov's eyes the social, political, economic, and technological conditions of Europe in the eighteenth century were different enough from the time at which he was writing to make obsolete the methods of late eighteenth-century warfare. One could only form a practicable strategy from studying the most recent wars,[93] as von Clausewitz had argued in his work, and for this Neelov was able in the latter part of the 1840s to restructure the study of military history to focus attention on the most recent military campaign while deemphasizing the study of ancient ones.[94]

In this conception of military power, Neelov's work became quite closely knit with Miliutin's *Attempt at a Military Statistics.* Neelov especially shared enthusiasm for the proper formulation of a strategic plan which was to take into account all the features of a theater of war for the purpose of systemizing the deployment and movement of troops against potential foes. The factors of a theater of war to be assessed in the creation of this plan were as follows: (1) political relations among the countries included in a particular theater of war; (2) the geographical and topographical characteristics of the theater of war; (3) the statistics or economic composition of the countries in the theater of war; and (4) the structure of the potential military establishments involved.[95]

Perhaps the most important structural concept of the strategic plan was to be that of Jomini's lines of operation, which provided a strong logistical link between the rear and the zone of military operations within the theater of war. In conjunction with Miliutin's recognition of the railroad's possible role in providing this stronger logistical support for military operations (a concept which did not come into use until its effectiveness was shown in the American Civil War in the 1860s as well as the Austro-Prussian War of 1866 and the Franco-Prussian War of 1870/71), Neelov tried to define more accurately its potential role in the formulating of the war plan in what was to be the first attempt to predict the requirements of future military operations.[96]

When Colonel Viemarn retired as head of the faculty of tactics in 1843, he was replaced by Colonel Goremykin, a graduate of the Nicholas Academy and one of the original contributors to the *Military Encyclopedic Lexicon.* Sharing the liberal orientation which was now predominant in the academy, Goremykin reiterated Medem's fundamental point of view that military science could not provide a "constant guide for action; only an indicator for the use of known truths in one or several cases."[97] He recognized that the military sciences were still very primitive but felt that as they were subjected to further research they would be able to provide an increasingly close tie between theory and art.

Goremykin's own research on tactics evolved throughout the 1840s, culminating in the publication of his *Guide to the Study of Tactics* (*Rukovodstvo k izucheniiu taktiki*) in 1849.[98] Within the specialized field of tactics, as in the study of strategy, the lesson to be learned was that means may remain comparatively constant within a short period of time and could be adequately studied, but that circumstances could encompass endless permutations. Any attempt to fully comprehend tactics would prove to be highly complex and limited in utility due to the field's highly specific nature. Likewise, the study of strategy as a science, or system of theories, could only provide a few very general principles about the conduct of military operations in the context of an entire campaign. Acknowledging the same attitude toward the nature of war which had been enunciated by von Clausewitz and which formed the basis of the liberal school, Goremykin concluded that

> finally, these few thoughts and elementary discussion which in tactics is accepted as the basis for rules is only applicable for [the present] time: born from experience they can only be applicable during that time when [the basis for this] experience has not changed; any new invention in weapons can change the arrangement and active abilities of the troops.[99]

Despite the repressive measures taken by the original members of the special commission and the director to mold the academy into a disciplinarian educational institution which would uphold service formalism of the autocracy, a glimmer of intellectual freedom promoted by Medem and maintained by his students became a permanent fixture in the life of the academy. Although the academy was not yet an official forum for the development of military philosophy in Russia as Jomini had originally intended, its instructors were permitted to review and evaluate many different hypotheses in light of Russian conditions.

NOTES

1. D. Reichel, "La position du Général Jomini en tant qu'expert militaire à la cour de Russie," *Actes du Symposium 1982*, Service historique, Travaux d'histoire militaire et de polémologie (Lausanne, 1982), 1: 64–65.
2. P. Geisman, "Uchrezhdenie imperatorskoi voennoi akademii," *Voennyi Sbornik* 11 (1908): 91.
3. John Shy, "Jomini," in *Makers of Modern Strategy: From Machiavelli to the Nuclear Age*, edited by P. Paret and M. Howard (Princeton: Princeton University Press, 1986), 84.
4. J. A. Gilbert, ed., *An Exposition of the First Principles of Grand Military Operations and Movements Compiled from the "Treatise upon Grand Military Operations" by the Baron de Jomini* (London: Whitehall, 1825), 144–45.
5. P. Geisman, "Uchrezhdenie Imperatorskoi voennoi akademii," *Voennyi sbornik* 12 (1908): 136.
6. Ibid., 11 (1908): 96.
7. Ibid., 97.
8. Ibid., 12 (1908): 137.
9. The formation of a commission to develop the official regulations and to determine

the character of the new higher military-educational institution was delayed because of the war with Turkey until Oct. 1829.

10. J. S. Curtiss, *The Russian Army under Nicholas I, 1825–1855* (Durham: Duke University Press, 1965), 106. In the founding year of the academy, twenty-seven students were enrolled, the number of which gradually dwindled to ten second-year and seven first-year students in 1851. However, with the comprehensive revamping of the academy following its transferral to the department of military-educational institutions in 1853, the enrollment figures increased significantly. Enrollment was limited to those graduates from the artillery and engineering schools who passed rigorous entrance examinations for the academy and fell far short of Jomini's concept of a large and diverse élite officer corps:

a staff, which should be good in every respect, should, in my opinion, be that which enjoys sufficient considerations and prerogatives, to offer advantages to the officers of all the other arms, and which would thus be composed of soldiers already distinguished by their fitness for the service. The officers of engineer and artillery would then cease to be the antagonists of an institution which would offer them a greater field of distinction, and which would thenceforth be a collection of those officers of these arms, the most capable of directing an operation of war, and put at the discretion of the commander-in-chief for his assistance. (Gilbert, *First Principles,* 145).

11. Geisman, "Uchrezhdenie Imperatorskoi voennoi akademii," 11 (1908): 95.

12. Ibid., 12 (1908): 14. See also N. P. Glinoetskii, *Istoricheskii ocherk Nikolaevskoi akademii general'nogo shtaba* (St. Petersburg, 1882), 54. The faculty of tactics and history of military art shared twelve lecture hours per week with the faculty of strategy, military history, and military literature; the faculty of military geography and statistics received eight lecture hours per week; the faculty of topography and geodesy received four lecture hours per week; and the subjects of Russian philology, artillery, fortifications, general staff duties, and foreign languages each received two lecture hours per week.

13. The lectures at the academy were given between 9 A.M. and 3 P.M., with two lectures lasting two hours each in the morning. There was a half-hour break before the third lecture.

14. Glinoetskii, *Istoricheskii ocherk Nikolaevskoi akademii general'nogo shtaba,* 28.

15. G. P. Meshcheriakov, *Russkaia voennaia mysl' v XIX v* (Moscow: Izdatel'stvo 'Nauka', 1973), 99.

16. Ibid., 112.

17. *Voennyi entsiklopedicheskyi lekhtsikon,* pod. red. L. I. Zeddeler (St. Petersburg, 1837–1849), 1: 454.

18. Meshcheriakov, *Russkaia voennaia mysl' v XIX v,* 89. See also Glinoetskii *Istoricheskii ocherk Nikolaevskoi akademii, general'nogo shtaba,* 106.

19. Glinoetskii, *Istoricheskii ocherk Nikolaevskoi akademii general'nogo shtaba,* 96.

20. Meshcheriakov, *Russkaia voennaia mysl' v XIX v,* 101.

21. Glinoetskii, *Istoricheskii ocherk Nikolaevskoi akademii general'nogo shtaba,* 97–100. See also N. S. Golitsyn, "Imperatorskaia voennaia Akademiia v 1834–48 gg.," *Russkaia starina* 27 (1883): 413. Although no longer part of the official forum of Russian military thought, Zeddeler still continued to make significant contributions to the development of military knowledge with the publication of his two-volume text, *Obozrenie istorii voennogo iskusstva* [*Review of the History of Military Art*], issued in 1836 and 1843, and his compilation of the influential *Voennyi entsiklopedicheskyi leksikon* [*Military Encyclopedic Lexicon*], published between 1837 and 1849, drawing upon the experience of many instructors at the Nicholas Academy.

22. Glinoetskii, *Istoricheskii ocherk Nikolaevskoi akademii general'nogo shtaba,* 110.

23. *Russkaia voenno-teoreticheskaia mysl' XIX i nachala XX vekov;* pod. red. L. G. Beskrovnogo (Moscow: Voennoe Izdatel'stvo Ministersiva Oborony Soiuza SSR., 1960), 102. This substantial collection by a great Soviet military historian contains a facsimile of Medem's essay, *Obozrenie izvestneishikh pravil i sistem strategii.* (Heretofore works from this compilation will be cited by their original titles.)

24. Medem, *Obozrenie izvestneishikh pravil i sistem strategii,* 99–102.

25. Glinoetskii, *Istoricheskii ocherk Nikolaevskoi akademii general'nogo shtaba,* 52.

26. Ibid. An example of the probable reading list suggested by Medem in the final segment of his course is provided under "Literature of Military History" in the *Voennyi entsiklopedicheskyi lekhtsikon* (13: 480–90).

27. Medem, *Obozrenie izvestneishikh pravil is sistem strategii,* 102.

28. Ibid., 115.

29. A. A. Strokov, *Istoriia voennaia iskusstva* (Moscow: Voenizdat, 1965), 283.

30. Ibid.

31. Medem, *Obozrenie izvestneishikh pravil i sistem strategii,* 115.

32. Strokov, *Istoriia voennaia iskusstva,* 286.

33. Ibid., 287.

34. N. Medem, *Taktika* (St. Petersburg, 1837), 86.

35. Ibid.

36. Ibid., 88–89.

37. *Voennyi entsiklopedicheskyi lekhtsikon,* 1: 454. Rudtorfer was the author of *Militairgeographie von Europa in Tabellen* (Vienna, 1835), which was translated into Russian two years later.

38. P. A. Yazykov, *Opyt teorii voennoi geografii* (St. Petersburg, 1838), 1: ii. See also Glinoetskii, *Istoricheskii ocherk Nikolaevskoi akademii general'nogo shtaba,* 59.

39. Glinoetskii, *Istoricheskii ocherk Nikolaevskoi akademii general'nogo shtaba,* 59.

40. Ibid., 60.

41. Yazykov, *Opyt teorii voennoi geografii,* 1:x. "on the other hand, I do not see the reason why we Russians must only repeat that which is said by foreign writers. It is not decreed by laws of nature that new ideas and discoveries in the sciences must without fail pass from the East to the West. They can go the other direction." Also cited in Strokov, *Istoriia voennaia iskusstva,* 288; and Peter Von Wahlde, "Military Thought in Imperial Russia" (Ph.D. diss., Indiana University, 1966), 66.

42. J. D. Hittle, ed., *Jomini and His "Summary of the Art of War,"* (Harrisburg, Pa.: Stackpole, 1965), 43.

43. Yazykov, *Opyt teorii voennoi geografii,* 18. This bears a similarity to Jomini's comment that

> under military policy we may include the passions of the people to be fought, their military system, their immediate means and reserves, their financial resources, the attachment they bear to their government or their institutions, the character of the executive, the characters and military abilities of the commanders of their armies, the influence of cabinet councils or councils of war at the capital upon their operations, the system of war favoured by their staff, the established force of the state and its armament, the military geography and statistics of the state which is to be invaded, and, finally, the resources and obstacles of every kind likely to be met. (Cited in Hittle, *Jomini and His "Summary of the Art of War,"* 53).

44. Hittle, *Jomini and His "Summary of the Art of War,"* 43.

45. Yazykov, *Opyt teorii voennoi geografii,* 19, 255.
46. Ibid., 25–26.
47. Ibid., 31.
48. Ibid., 33.
49. Ibid., 38. Yazykov's exact sentiments are reflected in the statement that "unfounded, that the configuration of the landscape can remain constant, it only remains constant for one historical epoche; but time without fail will change them. New passes over mountains, roads, towns, forts, and ports and borders: all these circumstances change the geographical aspect of the terrain."
50. Ibid., 43. His definition of natural geographical subjects included oceans, mountains, rivers, and physical features of terrain not created by humans; artful elements included ports, canals, roads, and anything which could be influenced by the activity of humans.
51. Ibid., 45.
52. Ibid., 51.
53. Ibid., 58.
54. Ibid.
55. Ibid., 217.
56. Proof of this can be seen in the fact that the article is not attributed to any author and that its description of the role and characteristics of military geography and statistics is quite different from Yazykov's concept: "Military Geography cannot be included as a science, although it is very helpful in the research of theatres of war and the composing of plans for strategic operations" (1: 453).
57. Meshcheriakov, *Russkaia voennaia mysl' v XIX v,* 113.
58. Golitsyn, "Imperatorskaia voennaia Akademiia v 1834–48 gg." 426. He did originally submit it to the academic council but they rejected it on grounds similar to Medem's reservations.
59. Yazykov, *Opyt teorii strategii* (St. Petersburg, 1842), 132: "General Medem, obviously, showed that each writer on strategy mainly based his discussion on one element; but all the same, they strive to discover general laws, and for that reason often enter into discussion, although of ill-defined and uncontrolled form, of influence on other elements." Also cited in Strokov, *Istoriia voennaia iskusstva,* 288.
60. Yazykov, *Opyt teorii strategii,* 132, 137.
61. Ibid., 139.
62. Ibid., 140.
63. Strokov, *Istoriia voennaia iskusstva* 291.
64. Yazykov, *Opyt teorii strategii,* 130.
65. Ibid., 142.
66. Strokov, *Istoriia voennaia iskusstva,* 291.
67. *Voennyi entsiklopedicheskyi leksikon* (1: 478) defines military history as follows:

> It looks at the reliable means to educate military people and to provide them with protection against false theories which sometimes one meets even in the best tactical, strategic, and other military essays. But in order to study Military History one needs to base oneself on accurate and prudent laws, and in order to have this military historical essays must be compiled by experienced people, fully knowledgeable of military affairs and devoid of bias and academic prejudices. Military history divides itself into two main parts: into History of Military Art and History of Wars. The first concerns the birth and development of Military Art from the beginning of warfare up to our time, and the reasons for its rise or fall; the second concerns itself with the description of military events. It is

unnecessary to note that for the study of either of these parts it is necessary to study the other: that without basic information in the composition of Military Art of a nation it is impossible to understand its wars, and vice-versa.

In his thesis on the early career of Miliutin, E. W. Brooks notes Miliutin's positive attitude toward the professional utility of this publication for the Russian army ("D. A. Miliutin: Life and Activity to 1856" [Ph.D. diss., Stanford University, 1970], 65).

68. Brooks, "D. A. Miliutin," 68.

69. Ibid.

70. Ibid., 113.

71. D. Miliutin, *Pervye opyty voennoi statistiki.* St. Petersburg, 1842, 168.

72. Ibid., 172–73. Cited also in Glinoetskii, *Istoricheskii ocherk Nikolaevskoi akademii general'nogo Shtaba,* 62.

73. Ibid., 177.

74. Ibid.

75. Ibid., 180.

76. L. G. Beskrovnyi, *Ocherki po istochnikovedeniu voennoi istorii Rossii* (Moscow: Izdatel'stvo Akademii Nauk SSSR, 1957), 288–89; Brooks, "D. A. Miliutin," 115. Criteria were to fall into two groups. The first, seeking general information, consisted of a general overview of a province in military-statistical terms, followed by information on the terrain, the inhabitants, the productivity of industry, agriculture and trade, and education. The second sought more specific information from the department of the general staff, the provisions departments, and the regional commissariats.

77. W. B. Lincoln, *Petr Petrovich Semenov-Tian-Shanskii: The Life of a Russian Geographer* (Newtonville, Mass., 1980), 54; Brooks, "D. A. Miliutin," 118–21. In a much later work by a twentieth-century Russian emigré historian, Miliutin's work on military statistics was considered to be the first truly military-scientific work in nineteenth-century European military literature (N. N. Golovin, "Sociology of War," Hoover Institition on War, Revolution, and Peace, N. N. Golovin Collection, box no. 11, 14).

78. Golitsyn, "Imperatorskaia voennaia Akademiia v 1834–48 gg.," 418. As students at the Nicholas Academy in the 1830s, Goremikin, Neelov, and Bogdanovich sat in the front row and furiously took notes as Medem lectured.

79. *Voennyi entsiklopedicheskyi leksikon,* 1: 383–84.

80. A. Kochetkov, "Russkii voennyi istorik M. I. Bogdanovich," *Voenno-istoricheskii zhurnal* 8 (1968): 83.

81. Ibid.

82. Ibid. Two of these works were *Zamechatel'neishie pokhody Petra Velikogo i Suvorova* (St. Petersburg, 1846); and *Zapiska strategii. Pravila vedeniia voiny, izvlechennye iz sochinenii Napoleona, ertsgertsoga Karla, generala Zhomini i drugikh voennykh pisatelei,* 2 vols. (St. Petersburg, 1847).

83. Ibid. The implication was that the theoretical influence of von Clausewitz's work on the development of Russian military theory was not felt until the late 1840s.

84. *Voennyi entsiklopedicheskyi leksikon,* 1: 384.

85. Kochetkov, "Russkii voennyi istornik M. I. Bogdanovich," 83.

86. *Voennyi entsiklopedicheskyi leksikon,* 1: 384 (emphasis added). This statement suggests that Bogdanovich did not deny the significance of material means for the conduct of warfare, only that the study of the development in material means on the conduct of warfare would not reveal the eternal and unconditional principles which would guide their

use. He thus concluded that such a study would not be useful to the development of a military science. It is also possible that Bogdanovich and Neelov had agreed to delegate the research accordingly: Bogdanovich concentrated on the subjective aspects of strategy, while Neelov dealt with the manipulation of material elements.

87. Although this maxim is often associated with von Clausewitz, it was not unfamiliar to the student of military theory in the 1820s and 1830s.

88. Kochetkov, "Russkii voennyi istornik M. I. Bogdanovich," 83.

89. Von Wahlde, "Military Thought in Imperial Russia," 74.

90. Ibid., 75.

91. Meshcheriakov, *Russkaia voennaia mysl' v XIX v,* 105.

92. Ibid., 118.

93. Ibid., 117.

94. Glinoetskii, *Istoricheskii ocherk Nikolaevskoi akademii general'nogo shtaba,* 55.

95. Meshcheriakov, *Russkaia voennaia mysl' v XIX v,* 119.

96. Ibid., 120.

97. *Voennyi entsiklopedicheskyi leksikon,* 1: 499.

98. F. Goremykin, *Rukovodstvo k izucheniiu taktiki Chast I. Nachal'naia taktika* (St. Petersburg, 1849).

99. Ibid., 193.

2

Revitalization: The Academy as a Military-Scientific Forum in an Era of Reform, 1847–1856

The two main components of the liberal school which evolved at the academy during the 1830s and 1840s, namely, the demystification of the cult of genius and reliance on statistics for improved policy over military issues, was to find parallel development among the intelligentsia and enlightened bureaucrats responsible for enacting state policy in the 1840s and early 1850s. Inspired by the need for policy not based on general rules of bureaucracy but on specific knowledge of real social and economic conditions gained from statistics, many of these educated bureaucrats were to develop a new ethic of civic responsibility within the Imperial Russian Geographical Society founded in 1845. That there should be two informal forums for the advancement of similar professional concerns was indicative of the pervasive crisis facing Russian society which eventually became manifest in the Crimean War (1853–1856): the unresponsiveness of Nicholas I's formalistic state bureaucracy to evolving social, economic, and military realities in the Russian Empire and Western Europe.

Originally founded under the auspices of Nicholas I for the purpose of supporting the seemingly innocuous applied science of geography, the Imperial Russian Geographical Society came under the sway of the enlightened bureaucrats in 1847 when they were able to wrest control of the society's program away from the originally appointed older generation of scientists known as the "Germans," who were preoccupied with foreign geographical studies. The new program of the society focused on the study of various aspects of geography pertaining to the Russian Empire, but after the swell of revolutionary fervor throughout Europe in 1848 which provoked a crack-down on liberal thought at home it became the only center of support for studies on the social and economic life of Russia.[1]

Among the young intellectuals who joined the society in the mid-1840s were Dmitrii Miliutin and his brother Nikolai, an ambitious diplomat in the ministry of interior, both of whom were to quickly become important activists for the new program. As a professor at the Nicholas Academy, Dmitrii was to contribute to various proposals, such as those of I. V. Khanykov on improving the scientific definition of Russian geographic terminology and of P. P. Semenov-Tian-Shanskii for a Russian geographical-statistical dictionary in 1850. He was to share his interest in Russian social and economic issues with many of the St. Petersburg intelligentsia, among whom were fellow lecturers M. I. Bogdanovich, G. F. Stefan, and A. P. Kartsov from the Nicholas Academy.

In the fall of 1848 Miliutin's expertise in late eighteenth-century Russian military history was to temporarily halt his work in the Geographical Society and duties as a lecturer at the academy., He completed an anthology of the war with France in 1799 which had been begun by the famous court historian, A. I. Mikhailovskii-Danilevskii. Keenly aware of the abuse of history as a vehicle for theory, Miliutin painstakingly studied extant historical documents over four years in order to present as accurately as possible Suvorov's last victorious campaigns against the French in Italy.[2] Well-received at court, the *History of the War in 1799 between Russia and France in the Reign of Paul I* (*Istoriia voiny 1799 goda mezhdu Rossiei i Frantsieiu v tsarstvovanie imperatora Pavla I*) was largely responsible for Miliutin's appointment as a corresponding member of the Imperial Academy of Sciences in 1853 and as a member of the academic council, a dual distinction unusual for a general staff officer.

Conversely, life at the Nicholas Academy was beginning to experience some changes for the better in the latter part of the 1840s. In September 1848 G. F. Stefan was called out of retirement to assume the post of vice-director which Colonel Rennenkampf had occupied upon Zeddeler's dismissal in 1834. Stefan had recommended Miliutin and Bogdanovich for posts as professors at the academy. He had been popular with the students when he gave lectures on military geography under Yazykov. His appointment as vice-director was an indication that the severe disciplinary control which Sukhozanet had held over the administrative and intellectual life of the academy was slackening. As a consequence, the debates in the academic conferences between Miliutin, Bogdanovich, Kartsov, and Lebedev (the last two instructors of tactics and military administration, respectively) over the structure and content of the curriculum became more lively and resulted in the reform of many of the courses.[3]

Sukhozanet's former draconian authority over the affairs of the academy was to be further eroded when the war minister admonished the academic council in 1851 over the appallingly few numbers of cadets enrolled or graduating from the academy.[4] The problem was not only the insufficient numbers of graduates from the academy; concern was also aired as to the effectiveness of the training. Observing in September 1851 that "today the courses at the Military Academy are too school-like, theoretical, and very pedantic; one must make it more practical and, for that, all the taught subjects must be directly linked with their practical use in the service,"[5] Miliutin was to draw up numerous memoranda proposing the increase of academic standards and more correlation between the abstract lecture material and practical exercises. The message was clear: major reforms were needed in the academy and throughout the higher military-educational institutions if they were to be brought closer to the academic needs of the cadet corps.[6]

When hostilities broke out between the Russian Empire and the French and British Entente powers over political influence in the Balkan region in 1853, the war ministry accelerated the process of reform in higher military education by appointing General Rostovtsev to chair the academic council and transferring administrative control of the academy into his hands. Rostovtsev immediately assigned the more active of the instructors into high administrative positions within the academy, making Goremykin the staff officer on duty and appointing Miliutin and Kartsov as chiefs of the academic sections.[7] The academic council also took this chance to reform the system of examination which had previously been highly exclusive, abandoning examination by oral recitation in favor of written examinations in preparation for the anticipated expansion of the student body.

Ultimately the reform of the higher military-educational system resulted in the subordination of the academy under the main chief of military educational institutions, the dismissal of Sukhozanet as director, and the appointment of vice-director Stefan as the new director of the academy in February 1854. Although this reformed higher military-educational system was able to graduate more officers fit for service, the losses sustained by the Russian army in the siege of Sevastopol demanded even more intense training of officers, resulting in the consolidation of the Nicholas Academy with the engineering and artillery academies in 1855. The combination of the academic conferences of the respective academies into one entity allowed the supervision of the curriculum and facilitated streamlining of the curriculum and monitoring of examinations.[8]

The Crimean War, precipitated by elaborate diplomatic miscalculations and misconceptions of political priorities among the major European powers over the religious and economic fate of the disintegrating Ottoman Empire, was the first major realignment in the balance of power in Europe following the Congress of Vienna in 1815. In response to the Emir Pasha's declaration of war in early 1853, Russia's armed forces quickly overran the Danubian principalities which, in turn, provoked French and British naval and military forces to enter the fray against the Russians along the littoral of the Black Sea. Because of the extended colonial empires of the four powers, the conflict between them took on an international scope with military engagements occurring in the Far East, the Caucasus and Trans-Caucasus, and the Baltic region.

Initially the Russian army and navy performed well on all fronts; but toward the end of 1854 with the successful Entente amphibious landing on the Crimean peninsula, Russian defenses at the port of Sevastopol became enmeshed in inconclusive siege operations against a technologically superior army. The Russian field command and members of Nicholas I's strategic planning commission were

ГЕНЕРАЛЪ-ЛЕЙТЕНАНТЪ

СТЕФАНЪ

Lieutenant-General Stefan

reluctant to supply the required troops from defense of the Baltic littoral and the Austrian border; in September 1855 Russian troops were forced to evacuate the defense of the port. Faced with the threat of Austrian military intervention on the side of the Entente in January 1856 and given the exhaustion of the Russian army, the new tsar, Alexander II, accepted the Austrian terms of peace.

When Nicholas I died in February 1855, he left to his successor a country still shackled to the limitations of the serf system and in the throes of defending native soil from the armies of the Entente. The loss of the fortress and port at Sevastopol to the Entente armies in the autumn of 1855 proved to be a signal that the Russian Empire, which had been a major power in European politics since the Congress of Vienna and had suffered under the oppressive centralism of Nicholas I's reign, could no longer compete with Western Europe economically, politically, or militarily. By the war's end in 1856 the empire's domestic situation had also taken a turn for the worse; it was threatened by both an extreme financial crisis exacerbated by the war and a growing revolutionary movement among peasants, intellectuals, and military professionals alike.

The poor performance of Nicholas I's highly formalized bureaucracy also invited attack from intellectual social activists whose articles in the anti-establishment press pointed to the defeat of the Russian garrison in Sevastopol as a manifestation of the fact that Russian society had become stagnant in comparison to the liberal political and industrialized economies in many Western European societies. More specifically, among many members of the military the poor performance of the Russian army was seen as an important indication that the conduct of warfare was changing in accordance with discoveries in science and the development of industrialized economies. In the words of a Soviet military historian reviewing this era, the awareness of the imperial military élite to "the necessary wide application of science to improve the military training of the officers and the development of their interests in military literature rose especially sharply after the Crimean War." The élite also recognized the need to reform the system of recruiting troops, to improve the logistics system, and to arm the troops with modern weapons.[9]

Criticisms of the administration of the army were made even during the Crimean War by the supreme commander of the guards and grenadier corps, General Ridiger.[10] In his official report to Alexander II on June 4, 1855, Ridiger complained that the poor performance of the imperial army was due to the absence of capable people in command and administrative positions. This was seen to be the result of two causes: overcentralization of the military administrative machinery which destroyed the personal qualities of the officers and a military-educational system which did not promote personal excellence. The net result was "a degeneration of personnel who had possessed military spirit and a knowledge of tactics and war to personnel who study only one military exercise and whose world-view is limited to regulations and parades."[11] To correct this state of affairs, Ridiger recommended a wide-ranging reform of the entire military establishment toward a decentralization of military administration as well as an improved military-educational system which would develop a higher level of leadership competence and initiative in contrast to the overly formalistic world of regulations and parades.

In a second note to the tsar on June 23, Ridiger recommended a series of proposals which would, in his opinion, address the serious problems inherent in the military system: (1) get rid of excessive centralization; (2) change the whole character of troop training; (3) rewrite all the war regulations; (4) raise the level of the officers' development (with more military education); (5) improve the more senior officer corps; and (6) discuss all questions in a special commission.

Although these suggestions represented a sharp attack against the existing official military traditions of the imperial army, they were accepted by Alexander II and a special Commission for the Improvement of the Military Element was established under the chairmanship of Ridiger on July 20, 1855.[12] Initially the members were limited to P. A. Dannenburg as vice-president, Lieutenant-Generals Maksimovich and Merkhelevich, Adjutant-General Prince Bariantinskii, Colonel Kartsov, Duke Baranov, and Major-Generals Barantsev, Glinka, Gechevich, Kurdiumov, and Loshkarev. Miliutin joined the commission in February 1856. The immediate tasks of the commission were the following: (1) to change and simplify the duty regulations; (2) to improve armaments; (3) to conduct exercises for the physical development of the troops and officers; (4) to create a new program for teaching troops in peacetime; (5) to change the system of officer attestation; and (6) to create new principles for the inclusion of officers in candidacy for the improvement of service.

Although Miliutin was not a member of the commission until February 1856, he was well aware of its work to decentralize the administration of the military establishment. A month later he presented a proposal entitled "Thoughts on the Disadvantageous Elements in the Military System in Russia and the Means by Which to Correct Them." Using a military-statistical analysis to compare the Russian armed forces with the effectiveness of those of other major European powers, Miliutin concluded that the main problem was the maintenance of a huge army in peacetime which led to the absence of trained reserves for deployment in wartime. The main reason for this seemed to lie in the nature of a society based on serfdom which did not allow a shortening of the length of service.[13] He suggested that this might be ameliorated by a change in the administration of troops to one based on a system of military districts. Although his proposals were not enthusiastically received due to the implications involved in such a large reorganization, Miliutin did not abandon his ideas for rationalizing the military administration. As chief of staff for the Caucasian army under Bariantinskii between 1856 and 1860, he was able to put some of his ideas into practice.[14]

Reforms specific to the education of field commanders and administrative staff were to receive confirmation in a statement from the inspectorate department in July 1856, which came to the conclusion that the officers graduating from the academy were lacking in proper training and knowledge of management and field duties. It recommended that these shortcomings be corrected by more practical exercises and acquaintance with all aspects of army affairs.[15] For the most part the military-educational community, especially the academy, reacted immediately and positively to these recommendations. They had found that attempts at fundamental reform of the curriculum in the past had not been possible due to overly conservative bias within the army, and were quite willing to reform the education of the student officers to conform with what they saw as a greater unity between theory and practice.

The first of the instructors to take advantage of the slackening of censorship to openly express his theoretical opinion on the direction of military affairs in the new era of industrial development was Colonel A. I. Astaf'ev. In *On Contemporary Military Art (O sovremennom voennom iskusstve)*, the first volume of which was published in 1856, Astaf'ev devoted much of his attention to the relation between technology and military art which had been so glaringly apparent in the conduct and outcome of the Crimean War. Acknowledging the influence of the great literary critic V. G. Belinskii on the development of his own theories on the nature of the military sciences, Astaf'ev was to intentionally fuel the polemic between the more progressive and reactionary interests in the imperial army by noting that "military art is a general affair requiring deep collegiate discussion and polemic: one man cannot do it justice."[16]

The main purpose of Astaf'ev's work was to somehow address the call for reform which was so urgently needed to cut the cost of maintaining an efficient military force. He understood that rationalization of military administration would not come of its own; rather, it would have to gain inspiration from a firm basis of military theory, a base which he set out to establish in his essay on the contemporary state of military art.[17] For Astaf'ev, the study of historical events was the most important key to the foundation of strategic theory and to all the military sciences, an opinion which he had derived from his predecessors at the academy and for which he paid special tribute to Medem. Conscious of the balance needed between subjective and empirical components in building a useful theory of military art, he shunned the purely speculative orientation which Bogdanovich had promoted before the commencement of the Crimean War. Astaf'ev was to give the study of the history of military art the highest priority. He felt that strategic truths existed but were hidden away in past military experience. These truths could be uncovered by the historical study of the actions of great leaders and the essays of military philosophers in both the classical and modern eras. Culled from the historical analysis of every nation's military experiences spanning all epochs of human history, these strategic truths consequently had a universal character and were simple to understand, providing a theoretical basis which would not change. However, technological development of armaments and the peculiarities of the political systems involved would greatly affect their successful application in the conduct of warfare. "Such discovered laws can change externally in accordance with the countless subjects and circumstances: but in essence they don't change their basis; they are like general algebraic formulas."[18]

Having explained the importance of the history of military art as the philosophical basis for a unified theory of strategy, Astaf'ev turned his attention to the description of the dialectic he felt existed in the nature of military knowledge. In complete agreement with the traditional liberal assumption that military knowledge had a dual nature consisting of material and moral factors, Astaf'ev was to clarify a trend in military thought which had up until then been confused: "The army consists of moral and material elements. It is assumed that the moral element is in the commander and the material element in the army as an organism at the call of its leader."[19] Medem believed that the role of the commander was to keep in touch with all the elements and changing circumstances – requiring deep consideration unhindered by limitations of static, geometrical theories. Astaf'ev related the moral aspect of the commander with a subjective approach to theory and implied that the material aspect embodied in the armament and manning of the army was best described by a positive, empirical approach, thus clearly defining the dual nature of military science.

However, the perfection of technology was to alter both the relationship between the commander and his army and the relationship between the subjective considerations of the commander and the empirical guidelines which were to regulate the activity of that army. This ever-fluctuating correlation between the two Astaf'ev illustrated in an analogy between the use of the bayonet and artillery (inspired by the recent historiography on Suvorov), pointing out that the perfection of artillery's effectiveness in battle demanded a restructuring of tactical organization toward extended order formation rather than the concentration of forces long postulated by the example of the Napoleonic wars: rearmament of the troops, a renewed significance in the use of terrain, and a swing of tactical advantage to the defensive.[20]

In trying to define this ever-fluctuating dualism in the military sciences, the primary concept to address was that of the relationship between "higher" and "lower" military art, or, for lack of more accurate terminology, strategy and tactics. Strategy, which was to synthesize all the factors influencing the conduct of war, was to provide to tactical manipulation of troops the conditions upon which to base its fighting order and regulations because, as Astaf'ev pointed out, "battle is part of war."[21] Even the concept of tactics could be typologized into lower tactics, concerned with the correlation of tactical formations within the circumstances of the battlefield and perfected by drill and field exercises, and higher tactics, concerned with the coalescing of the commander's mind with the troops' actual performance in battle. Both typologies were an attempt to accurately define that point of contact between the empirical and the subjective, the army and the commander, which formed the fundamental paradox within the military profession.

With the dualism in military affairs fairly well established, Astaf'ev turned his attention to the definition of military science. Like his predecessors, he was sympathetic to the notion that the art of strategy would assume scientific properties as it developed. In his opinion, a science was needed to facilitate the evolving relationship between the nation and the army as the conduct of warfare involved more and more of the nation's resources. However, until this time the concept of military science had been loosely defined as the code of theory concerned with the whole range of subjects related to military affairs. Astaf'ev did not fundamentally disagree with this traditional approach but found it lacking in accuracy and therefore of little positive practical use. In an attempt to provide a clearer and more useful definition, he developed a typology to provide the parameters and internal structure of military science: tactics, artillery, engineering, military administration, military geography and statistics, topography, strategy and military history.[22] These would be the basic, or in his words, the "atomic" sciences which would provide the best description of the activities of an army in war for the attainment of victory. Research of military science would be conducted along much the same lines as research methodologies suggested by Medem, Yazykov, and Miliutin: a review of contemporary theories of strategic systems to examine all the strategic elements and the interrelationships between them, and an accurate definition of the influence each strategic element had on the conduct of the army in war.

A third aspect of past research methodologies had emphasized the synthesis of these strategic elements into a unified theory, an endeavor which Astaf'ev felt was completely unfounded and which would inevitably result in chaos.[23] One simply could not mix these dual components of military theory together to form a unified theory as previous theories had tried to do. The vastness of military science required a simultaneous differentiation and integration of abstract, subjective considerations with material factors and circumstances having direct influence on the activities of the army. As an illustration of this differentiation/integration process within military science, Astaf'ev posed this question: What does circumstance demand from the composition, organization, and order of the army (its administration) and how are these factors used to secure the attainment of political goals by success in war (military art)? The differentiation/integration process of military science also had significance for the role the auxiliary or non-atomic sciences played within military science as a whole. The dual nature or combination of subjective with material aspects of military science allowed the commander to benefit from the advantages of these speculative sciences without combining (and confusing) them with the empirical evaluation of military operations.

Having thus explained the nature of military science, Astaf'ev drew upon the implications of this military science to form the basis of reforms within all military institutions of the imperial Russian army. The first use of this theory of military science was to build a model for the ideal army as a living organism, not a machine carrying out orders—a model which must have gained inspiration from Ridiger's criticism of the imperial army's performance in the Crimean War and the inspectorate department's statement concerning the poor training of general staff officers for field service.[24] In contemplating the model for the ideal army, Astaf'ev felt that the commander should have the benefit of a universal education, military training over a long period of time, full knowledge of the material and capabilities of his army, and a sophisticated knowledge of current military art. Because the general staff provided an important link between the commander and his army, these officers should also receive a universal education to enhance awareness of their troops and a familiarity with every type of service.[25] In discussing the duties of the general staff, Astaf'ev distinguished between its duties in peace and war, and was quick to note that Jomini's emphasis on logistics as the special science of the general staff was too limited. The general staff was to assume a managerial role and thus benefit from the study of military geography/statistics and military administration.

Astaf'ev was especially interested in the topic of military administration. Although it was an essential component of military science, it had up to that time received very little attention. The laws of military administration were to be taken from the general subject of government administration because military administration was merely a subset of the entire state apparatus and military affairs were organically united with the economic condition of the state itself.[26] Naturally enough, Astaf'ev regarded military administration as an empirical science for the reason that it concerned itself entirely with the material aspect of military affairs. He suggested that the regulations pertaining to the administration of a corps could provide the scientific basis for other units, thus forming a link with tactical formations.

In general, the laws of military administration in both war and peace were to have as their main goal the preservation of the troops' composition and logistical support and the perfection of each unit making up the organization of the army. Astaf'ev felt the need to break this subject down into two levels of operation, the higher one belonging to the domain of the war ministry and the lower one to a smaller military-geographical-administrative entity, presumably that of the military district which was Miliutin's as yet nonexistent administrative system.[27]

> Lower military administration must concern itself with everything having to do with the individual soldier: from his weapon to the teaching of ideas necessary to him; to give laws to his education which he must strive to attain in order to come closer to perfection, the attainment of those goals set by his commander's evaluation.[28]

This was a point of view which he shared with Miliutin and suggested the increasing importance of political as well as operational instruction.

The experience of the Crimean War was to have significant influence upon other military theoreticians at the academy as well, especially Bogdanovich. By 1854, the year when his two articles, "About the Newly Perfected Hand Firearms" and "On the Influence of the Technical Aspects of Military Affairs on Tactics," were published in *Voennyi zhurnal*, Bogdanovich's former confidence in the existence of a positive, unified theory of strategy based on the moral characteristics of the commander had been severely shaken. When his article, "On the Study of Military History," appeared a year later it was obvious that Bogdanovich's orientation to the nature of strategic theory had come full circle and now expressed itself in the statement that "in military affairs nothing is indisputable. Insufficiencies of supply and bad weather can have primary importance on the outcome of actions"—not a very sophisticated reassessment, but a step in a different direction nonetheless.[29]

Bogdanovich's next major work, *History of the Patriot War of 1812* (*Istoriia Otechestvennoi voiny 1812, goda, po dostovernym istochnikam*), published in 1859/60 in anticipation of the fiftieth anniversary of the Patriotic War, received much attention because it was a strange amalgam of old methodological priorities with attempts to address the new ones which had emerged from his reactions to the Crimean War. Bogdanovich felt that the importance of the War of 1812 was its relative proximity in time and that an exposition of the activities of great generals would always retain its pedagogical significance. He also based his work, as was the case throughout his career, on the maximum use of documentary material. He departed from his previous methodology, however, by devoting increased attention to the material aspects which could influence the conduct of war, in particular the geographical and meteorological influences as well as the logistical support of the armies. This work was to receive much praise and an enormous amount of criticism from the general press. The main critic to voice his reservations was I. P. Liprandi, who felt that history should address the place and role of the nation in history and that Bogdanovich's work had failed to meet these broad goals.[30] The aspiring novelist Tolstoy also took part in the fray in his novel *War and Peace,* which he was writing at the time, by criticizing Bogdanovich's poor estimation of Kutuzov's role as commander of Russian forces against Napoleon's invading army and his faulty historical methodology.[31]

Although the end of the war in 1856 found the financial state of the Russian Empire in shambles, pressure from the Ridiger Commission and a high degree of sympathy on the part of the new tsar, Alexander II, for the training of more general staff officers fully competent in the management of the army resulted in a significant increase in the enrollment of student officers at the academy, a tremendous growth in financial resources, and the need for more instructors. The consolidation of the higher military-educational institutions into one entity in 1853 had stimulated the gradual growth in incoming first-year students which was to continue throughout the duration of the war. As a result of this increasing burden on the teaching facilities at the academy, the examination system was reformed under the auspices of Rostovtsev in 1854 and supplemented a year later by the formal establishment of student competitions. The best research projects were published in the *Academic Digest* of the academy.[32]

Faced with the potential problems a rapidly expanding student body might cause, the director, G. F. Stefan, proposed that the academic conference address the need to impose new criteria for hiring instructors. A special commission, consisting of Bogdanovich, Kartsov, Lebedev, and Maksheev, compiled rules for examining prospective professors, adjunct-professors, and instructors at the academy in an attempt to increase, or at the very least, maintain academic standards in the face of increased teaching demand.[33] Adopted on October 10, 1856, the new rules stipulated that in applying to teach at the more important faculties of tactics, strategy and military history, military administration, and military statistics/geography, the prospective applicant had to first pass an examination, then write and defend a dissertation. It was not until 1859/60 that the first dissertations were defended. One of the first was defended by A. E. Stankevich, who subsequently received a position as adjunct-professor in 1860.[34] The increased financial advantages of being subordinate to the department of higher military-educational institutions as well as the recognized need to study different military systems prompted Rostovtsev to permit the academic conference to select one of its academic personnel for study abroad. The first instructor to enjoy this new program was V. M. Anichkov, a protégé of Miliutin who had studied the history of the Caucasus. In 1858 M. I. Dragomirov was sent to study tactics and came back a year later to teach this subject in the academy. In 1860 two more protégés of Miliutin, N. N. Obruchev as professor of military statistics and N. P. Glinoetskii as adjunct-professor of military statistics, were sent to France and the Caucasus, respectively.

Rapid growth of the student body posed other problems for the academy as well. The increased number of students necessitated a physical reform of the course structure in 1857 which divided the theoretical and practical sections into two subsections. However, this was not sufficient; the curriculum was to gradually undergo a reform in content as well in response to the call for general staff officers of higher practical and managerial training.

The faculty to undergo the most change was that of tactics. From the early 1850s this course had been taught by Colonel A. P. Kartsov, a close friend of Miliutin who shared the prevailing predilection for a broadly humanistic educational curriculum at the academy.[35] Relinquishing his lectureship in 1856 to tutor the young heir apparent, he was replaced by Colonel Lebedev, a former student of the academy who was also responsible for teaching the course on military administration.[36] This course, a subject thrown into the floodlights of attention partly by Astaf'ev's controversial essay and partly by the very real financial, moral, and operational crises which were paralyzing the imperial army, could not be properly taught or researched by Lebedev. Only when it was taken over by Captain Anichkov in 1857 did it receive adequate theoretical development.[37]

Relief from teaching both courses simultaneously and without assistance was to come in 1858 when G. A. Leer, then M. I. Dragomirov two years later, joined the faculty of tactics. From the beginning of their work together each one specialized in an aspect of military art: Dragomirov was concerned with the training of troops whereas Leer tended to focus on methodological issues and the study of grand tactics (strategy), eventually becoming the professor of strategy in 1865.[38] During the early 1860s both intellectually ambitious men—who were to become monolithic figures in Russian military thought throughout the remainder of the nineteenth century—performed an extensive analysis of the Crimean War, redirected the course of Russian military art, and with Obruchev's introduction of field exercises and war games acquired during his assignment in France, increased the sophistication of tactics.

During Sukhozanet's years as director of the academy and with incorporation of analysis of the Crimean War, the course on military history under Bogdanovich had gradually expanded in content until by 1860 it was completely unmanageable. Only in 1861/62 at the instigation of the new director of the academy, A. K. Baumgarten, was the course on military history and theory of strategy reevaluated and abridged. In the theoretical section of the course the history of military art now held sway after having suffered under the tutelage of the faculty of tactics and being burdened by a disproportionate focus on ancient military art.[39] The practical section, on the other hand, consisted of discussion and written exercises on the theory of strategy and a review of the history of wars (different from the history of military art) as a mechanism to study the manifestation of military art and to illustrate the mistakes made by commanders and staff officers.[40]

When Miliutin left his post as professor of military statistics in the spring of 1856 for a position on the Ridiger Commission and subsequently for the post of chief of staff of the Caucasus army, his place was filled by Maksheev and Obruchev.[41] For the most part the fundamental structure of Miliutin's lecture remained unchanged, but Maksheev, in an attempt to cover the study of Russia's offensive as well as defensive military might, substantially enlarged the content of the course. Maksheev was responsible for the teaching of military statistics as a scientific methodology and as a component of military administration. Both these courses were now merged as a consequence of the emerging role of the general staff in the management of the Russian imperial army and Astaf'ev's work. Obruchev was responsible for providing a review of the military power of foreign nations.[42]

The turbulent years between the commencement of the Crimean War and the emancipation of the serfs in the early 1860s were highly productive ones for the development of military thought in the Nicholas Academy. The work of Miliutin and Astaf'ev during these years was to provide the basis for extensive organizational reform of military institutions in the following decade, and the revitalized intellectual climate within the academy was to produce new theoreticians who were to dominate the development of Russian military thought up to the First World War.

NOTES

1. W. B. Lincoln, *Petr Petrovich Semenov-Tian-Shanskii: The Life of a Russian Geographer* (Newtonville, Mass., 1980), 54; E. W. Brooks, "D. A. Miliutin: Life and Activity to 1856" (Ph.D. diss., Stanford University, 1970), 118–21.
2. Brooks, "D. A. Miliutin," 128.
3. N. P. Glinoetskii, *Istoricheskii ocherk Nikolaevskoi akademii general'nogo shtaba* (St. Petersburg, 1882), 101.

4. Ibid., 104.

5. P. A. Zaionchkovskii, *Voennye reformy 1860–1870 godov v Rossii* (Moscow: Izdatel'stvo Moskovskogo Universiteta, 1952), 231. This point was to receive significant attention when Miliutin as war minister addressed the need to reform the military-educational system in 1865.

6. Glinoetskii, *Istoricheskii ocherk Nikolaevskii akademii general'nogo shtaba,* 134.

7. Ibid., 126–27.

8. Ibid., 129, 153, 156.

9. G. P. Meshcheriakov, *Russkaia voennaia mysl' v XIX v* (Moscow: Izdatel'stvo 'Nauka', 1973), 162.

10. Zaionchkovskii, *Voennye reformy 1860–1870 godov v Rossii,* 45.

11. Ibid.

12. Ibid.

13. Ibid., 50. See also L. G. Beskrovnyi, *Russkaia armiia i flot v XIX veke. Voenno-ekonomicheskii potentsial Rossii* (Moscow: Izdal'stvo 'Nauka', 1973), 205.

14. Ibid., 84–85.

15. Glinoetskii, *Istoricheskii ocherk Nikolaevskoi akademii general'nogo shtaba,* 138.

16. Meshcheriakov, *Russkaia voennaia mysl' v XIX v,* 87. Also cited in A. Astaf'ev, *O sovremennom voennom iskusstve. Chast' I-II* (St. Petersburg, 1856–1862), 1:204; Beskrovnyi, *Russkaia armiia i flot v XIX veke,* 375; and A. A. Strokov, *Istoriia voennaia iskusstva* (Moscow: Voenizdat, 1965), 609.

17. Astaf'ev, *O sovremennom voennom iskusstve,* 203: "Reform in its material organization must be done after examination of advantages and disadvantages of innovations expressed in the laws of military science."

18. Ibid., 234.

19. Ibid., 204.

20. Ibid., 245. In explanation of what he meant by this swing of tactical advantage to the defensive is the following statement: "Today the destructive power of artillery returns the advantage of fortified positions, only possible assaults to be accomplished by quick attacks and enveloping/flank manoeuvres to take advantage of weaker sides of the fortification." Also cited in Strokov, *Istoriia voennaia iskusstva,* 611.

21. Ibid., 246.

22. Ibid., 209.

23. Ibid., 232–33: "Today one accepts the belief that the theory of strategy cannot create any purely speculative truths, but must be limited to the temporary subordination of chance which one finds in war, and to show the conclusions in sequential order from the activities of great commanders. Many hypotheses and limited strategies still exist which often contradict each other. Its theory presents a chaos of confused ideas."

24. Ibid., 206. Also cited in Strokov, *Istoriia voennaia iskusstva,* 612.

25. Ibid., 207. Astaf'ev's view of the general staff was identical to that expressed by Jomini thirty years earlier: "From these restriction [sic] one must exclude the officers of the General Staff as executers of the supreme commander's ideas; they must have a well-rounded education upon which he can rely because they are the special link between the commander and the army; they must transmit his will and concepts."

26. Ibid., 240.

27. Ibid., 241. Astaf'ev gives a list of all the institutions responsible for administration of the army on p. 242.

28. Ibid., 241.

29. A. Kochetkov, "Russkii voennyi istornik M. I. Bogdanovich," *Voenno-istoricheskyi zhurnal* 8 (1968): 83.

30. Ibid., 79.

31. L. Tolstoy, *War and Peace*, 4 vols. (London, 1915). Tolstoy makes a direct reference to Bogdanovich's work (4: 195). One of his many comments on military theory is as follows:

> Military science, seeing in history an infinite collection of examples, that the mass of armies does not coincide with the strength, and that small detachments have conquered large ones, confusedly recognises the existence of this unknown factor (x), and tries to discover it now in geometrical combinations, now in differences of armament, now, and this most generally, in the genius of the commanders. . . . This x is the spirit of the army; in other words, the more or less intense desire of all the men composing the army to fight and expose themselves to perils, independently of the question whether they are armed with clubs, or with guns delivering thirty shots a minute. . . . The spirit of the army is the factor, multiplied by the mass, which gives the product, power. To determine and express the meaning of the spirit of the army—that unknown factor—is the problem of science. (2: 13.)

32. Glinoetskii, *Istoricheskii ocherk Nikolaevskoi akademii general'nogo shtaba*, 186.

33. Ibid., 143.

34. Ibid., 145.

35. E. W. Brooks, "The Improbable Connection: D. A. Miliutin and N. G. Cernysevskij, 1848–1862," *Jahrbucher fur Geschichte Osteuropas* 37 (1989): 34–35.

36. Glinoetskii, *Istoricheskii ocherk Nikolaevskoi akademii general'nogo shtaba*, 166.

37. For more information about the military academic career of Anichkov, see Brooks, "Improbable Connection," 35.

38. Glinoetskii, *Istoricheskii ocherk Nikolaevskoi akademii general'nogo shtaba*, 160.

39. Ibid., 161. Neelov had tried to cut this out in his course on the history of military art in the mid- to late 1840s.

40. Strokov, *Istoriia voennaia iskusstva*, 625.

41. Glinoetskii, *Istoricheskii ocherk Nikolaevskoi akademii, general'nogo shtaba*, 163.

42. Ibid., 164.

3

Transformation: Turning Theory into Practice in the Miliutin Era, 1856–1878

The coronation of the new tsar and the cessation of hostilities with the Allied powers by early 1856 were welcomed by many social groups in the Russian Empire as the beginning of a new era of social reform and political centralization. In particular the loosening of government censorship over the press and the official approval of the Ridiger Commission to examine the reform of the Russian army were met with great enthusiasm by the members of the literary and military intelligentsia who were concerned with the social stagnation and waning military power of the Russian Empire. However, the first four years of Alexander II's reign were marked by indecision over how to overcome these acute problems. The tsar's attempts to balance conservative and reform-oriented interests led to vacillations in policy which were only to frustrate modernization efforts.

Throughout the previous decade the liberal social philosophy of such men as Alexander Herzen, N. S. Chernyshevskii, N. A. Dobroliubov, N. P. Ogarev, and others had permeated deeply into enlightened Russian society, especially among the members of the military intelligentsia interested in the preservation of the Russian Empire's military integrity. Perhaps the most influential of this work was the literary criticism of V. G. Belinskii, which inspired such reform-oriented military theorists as Astaf'ev to comment that

> for the time being philosophical debate was not resolved as to what was the best way to seek the truth in the sciences—*a priori* or *a posteriori;* we accepted Belinskii's thoughts on Truth and were to be guided by them in the creation of an ideal army and in the research of the exact significance of the essence of military science.[1]

This blending of social and military interests was particularly characteristic of N. S. Chernishevskii, who had been an instructor of literature and Russian philology at the Second Cadet Corps during the Crimean War. Although he left that post in 1856, he had retained close social ties with many of his friends from the Nicholas Academy and the artillery academy, including Miliutin, Obruchev, Kartsov, and Lavrov.[2] When Miliutin's suggestion for a new, uncensored forum for military issues was approved and the journal *Voennyi sbornik* was established in 1858, Chernyshevskii was appointed as principal literary editor, supported by N. N. Obruchev and V. M. Anichkov as editors for the military affairs section. For the first eight months of this journal, which was to later become the official mouthpiece of the war ministry, these three men adopted a highly critical tone toward the overly centralized rule of Nicholas I and the poor performance of his generals in the Crimean War.[3]

Although Chernyshevskii was soon to become a victim of opposing factions within the war ministry and was forced to give up his position as literary editor to his rival, General Men'kov, his brief tenure advanced the cause for a broader humanistic ethic throughout the entire Russian military establishment. Later, as the hopes of Chernyshevskii and other advocates for social change in the Russian Empire turned to revolutionary socialism, the effect of these ideas on the already smoldering revolutionary sentiments within the cadet corps took its toll. With mounting frustration and the founding of secret, politically subversive societies the aspirations of the officer corps coalesced around these ideas, epitomized in a cadet's observation that "we read and reread Dobroliubov's works but Chernyshevskii had a stronger and more direct influence with us. We memorized his works by heart, we swore by him."[4]

For the time being, however, the enthusiasm of these enlightened members of the social and military intelligentsia was frequently misled by Alexander II's enactment of policy which seemed to directly contradict his initial movements toward reform. In April 1856 Alexander II appointed General N. O. Sukhozanet, brother of the former director of the Nicholas Academy, as the new war minister to replace Prince V. A. Dolgorukov who had held that position during the Crimean War. Intent on ameliorating the impending budgetary and organizational crises in the war ministry by administrative adjustment rather than widescale reform, he sought a loyal bureaucrat who would respond to royal prerogative. Although Sukhozanet was able to enact his sovereign's specific reforms by cutting some of the administrative fat from the institutional responsibilities of the war ministry, most of the more substantial budgetary cuts merely depleted strategic stockpiles. Nor was he particularly attuned to enacting the more substantial organizational and educational reforms which had been suggested by the tsar and the Ridiger Commission.[5]

The incompatibility of desperately needed reform with the war minister's traditional formalistic attitude toward the function of the imperial army did not come to the fore until the Franco-Austrian War in 1859. When Napoleon III requested that the Russian army threaten the Austrians with the mobilization of four army corps to the Austro-Russian border, the war ministry was not capable of mobilizing the promised forces in less than five months. The ministry was thus shown to be deeply suffering from administrative stagnation.

When the semi-autonomous Caucasian army conducted several successful military campaigns in 1859 its outspoken commander, Prince Bariantinskii, gained a favorable position at court to propose his own ideas for substantial reform. These consisted of the experiments with territorial administration which he and his chief of staff, General D. A. Miliutin, had used against the mountain brigands in the Caucasus. Encouraged by support for his ideas at court, Bariantinskii proposed that the empire's military forces be placed under a similar territorial administration and suggested that the chief of the general staff, responsible for operational concerns, be elevated to the same political level of access to the tsar. The war ministry was to be reduced to merely a general administrative function. Successfully convincing the tsar that this model was advantageous, Bariantinskii was appointed chief of the general staff and his former chief of staff was appointed to the post of deputy to the war minister in August 1860 to eventually become the war minister in November 1861.

Although Miliutin had in the past shared in the general concern for the social welfare of the nation, as an enlightened bureaucrat seeking reform in the controlled setting of consultative committees and editing commissions he baulked at the argument for unbridled violence of popular revolution advocated by the increasingly revolutionary socialists.[6] Upon his appointment to the war ministry, Miliutin was to react to the protests of the students at the Nicholas Engineering Academy and various other disturbances from the position that

> the cadet corps, the infantry school, and even the three academies during the last year [1862] have issued into the troops young officers with such false and harmful ideas that their evil influence had revealed itself, in the opinion of the supreme command, everywhere in the troops and especially in the special type of arms, with the exception of the cavalry.[7]

In such a way were the liberal motivations of Miliutin and Obruchev, and subsequently the rest of the military intelligentsia, to break away from the Decembrist tradition advocated by the revolutionary socialists and attempt to promote gradual political and social change of the Russian nation toward a more modern European model.

Miliutin wasted no time in coming to terms with the enormous responsibilities of reforming the machinations of the war ministry. In an attempt to consolidate his position and guarantee the efficient execution of his proposals, he spent the first two months appointing new people to important administrative posts within the war ministry. He then began the arduous task of institutional reform by establishing several special commissions to study the nature of the problems facing the war ministry and their successful resolution. One in particular was the commission under the incumbent director of the Nicholas Academy, General Baumgar-

Adjutant-General Count Miliutin

ten, to address questions of the army's reorganization. Within two months of their establishment these various special commissions, meeting daily with the war minister and his aides, were able to formulate a *vsepoddaneishyi doklad* or general proposal of reform to be presented to the tsar on January 15, 1862.[8]

Reflecting many of the ideas Miliutin had proposed to the Ridiger Commission and the various academic councils throughout the 1850s, the *vsepoddanneishyi doklad* was to reform the armed forces of the empire in order to support the autocracy in an era of rapid social and economic development. The proposal recognized that the task at hand was a very difficult one due to the many contradictory goals: "On one hand it must seek to the utmost to ease those times when military expenses mount for the government, on the other hand it provokes reproach if it only worries about the cutting of budgets to the detriment of a well composed and well organized army."[9]

Miliutin was well aware of the importance of a strong army for the political strength of the Russian Empire in the European arena and Tsarist Poland, but stressed that it was also necessary for internal security to counter the growth of domestic revolutionary movements at the beginning of the 1860s. Confronted with these tasks he concluded that the Russian army needed to be regulated by a more rational military system than that existing in various European armies: Russia needed to decrease the size of the army in time of peace and efficiently increase its size in time of war in order to cut state expenditure. This type of system depended on the proper organization of reserves which would not be possible under the system of conscription and long periods of conscript-service. The main aim of the reform was meant to

> significantly shorten the available number of troops in peacetime with another no less important condition, the preparation of all means for the development of all arms in the case of war. It is necessary today to seek measures which prepare in leisurely fashion military strength of the Empire in personnel as well as material toward as correct as possible cheap mobilization and, establishing together with that a way of rational economy in annual military expenses, to protect in those moments the government capacity without sudden destructive shock to conduct war according to its political goals. This is the goal for which the War Ministry strives.[10]

With the nature of the challenge thus defined, the *vsepoddanneishyi doklad* was to meet it by fusing central administrative institutions in the war ministry with the hope that such a measure would cut out redundant, non-combatant elements. A significant amount of the war ministry's administrative load was also to be transferred to permanent regional organs of military administration which would assume regional domestic and national security responsibilities—the future military district system.

Having established two major programs of reform, Miliutin turned his attention to the central institutional anatomy of the war ministry itself. He immediately placed the artillery and engineering troops under the jurisdiction of the war minister within their respective main administrations. He then fused the provisions department and the commissariat into the intendancy administration. A year later, Miliutin was given permission to transfer the military-educational institutions to the domain of the war ministry and then, with the intention of unifying all the institutions concerned with general staff work within the war ministry, he combined the department of the general staff with the military-topographical section and the Nicholas Academy to form the main administration – a consolidation of power within the war ministry which was directly antagonistic to Bariantinskii's original intentions. However, as the reform of troop organization and the creation of the military district system progressed, the main administration of the general staff was once again forced to conform to change and was consolidated with the inspectorate department to form the main staff in 1866.[11]

In anticipation of the creation of the administrative military district system, the organization of the troops had to be resolved. The experience of the Crimean War had shown that the large tactical unit structure of the corps was not conducive to quick mobilization from a peacetime to wartime footing.[12] Gaining inspiration from the French army's success with the proven flexibility of the division as the largest tactical unit, Miliutin in his *vsepoddanneishyi doklad* proposed that the corps-army structure be changed in favor of a regiment-battalion-company system under the jurisdiction of the regional military districts and that these regiments would in peacetime (as well as in wartime) be divided into active and reserve. A regiment within an active or frontier military district would have a four-battalion structure, whereas the reserve regiment in an internal military district would have a two-battalion system.[13]

In September 1862 a special committee consisting of thirty-one members under the chairmanship of General Dannenburg was established to examine the proposed changes in troop organization.[14] Miliutin's proposals were to receive detailed discussion in this committee and immediately attracted opposition from those higher ranking commanders whose status would be put in jeopardy with the abolition of the corps structure. Arguing in favor of the corps system and against the proposed military district system, they focused upon the exact definition of the military district commander's powers as the issue of contention.[15] However, Miliutin was finally able to gain the upper hand over the opposition in this debate in the summer of 1864, when tensions with Western Europe over the two-year suppression of the Polish uprising abated. In August of that year Alexander II approved Miliutin's thesis on reserve troops.[16]

As regards the territorial military district system itself, Miliutin's *vsepoddanneishyi doklad* proposed the formation of local administrative bodies which were to play an integral part in the reform of the central military apparatus. He had originally proposed it in 1856, while a member of the Ridiger Commission, initiating the topic with the observation that a detachment staff was more suitable to the potential demands of a theater of war than the more cumbersome and frequently incomplete corps. At that time he pointed out that a territorial system of military administration in the form of military districts could provide the structure for such a detachment staff and could free the war ministry from many second-level administrative functions as well as provide a structure well suited for domestic security.[17]

At the beginning of May 1862 Miliutin presented to Alexander II a more detailed proposal devoted exclusively to the creation of the military district system. He once again stressed the need for this territorial-administrative system:

> The experience of many wars has shown us that the corps structure is too large a tactical entity for continued use in the Theater of War in all its complement; that almost always in different strategic considerations in the offence time must be taken to form detachments from troops of different corps to make special "detachment staffs"; all this could be avoided if it is ready-built into the administrative system.[18]

This reform could be brought about if the traditional organization of troops into corps and armies were to be abolished in favor of the division as the highest tactical entity. The empire's territory could be divided up into fifteen military districts according to military, statistical, ethnographic, territorial, administrative, and economic considerations; an administration consisting of a senior commander presiding over a district council as well as commanding the troops within the jurisdiction of that particular military district could be established.[19] The advantage of this system, in Miliutin's opinion, was that the applicability of such a system had already been demonstrated by similar entities already existing in the more remote areas of the empire.

In the middle of May 1862 this proposal was distributed to the military élite for comments which were to be taken into consideration in the drafting of an official thesis to be published in August 1862. However, due to the Polish uprising in the summer of 1862, Miliutin was able to force through the immediate creation of three military districts on the western border: the Warsaw, Vilna, and Kiev districts. The Odessa district was established a year later without taking into account criticism and recommendations from the military establishment.[20] During this time Miliutin's military district system received a large amount of criticism, as well as approval, from the officer corps of the imperial army.

After a delay of six months two editorial commissions were created to begin collating material for the official thesis on the military district system. The first commission was concerned with questions of command structure. It was presided over by General Geiden and had three other members (Shtang, Yakimovich, and Obruchev). The second commission on administrative-economic issues was chaired by Ustrialov, Polivanov, Piktorov, and Anichkov. After this material was collected, three more editorial commissions were established to look into issues concerning artillery, engineering, and military-medical affairs. By August 1864 this thesis on the military districts was approved and the four existing military districts were gradually augmented by another eleven for a total of fifteen by 1869.[21]

Despite the enormous amount of criticism and bureaucratic review to which

Miliutin's original proposals were subjected, the final structure of the military district system closely resembled that which he had proposed to the Ridiger Commission in 1856: a system of internal and frontier military districts to provide active and reserve forces, respectively, with the senior commander of the military district presiding over the administrative district council. This was a structure which would allow the peacetime military district administration to directly assume the administrative duties of a field army in time of war—a final responsibility which found full expression in the new regulations on field administration issued in 1868.[22] Another positive benefit of the military district system was the rationalization of the logistics system. No longer would the task of supply and logistic support be left to the discretion of individual commanders of field armies; it now became the concern of the stationary military district councils and intendancies. Under the rationalized logistics system a division in the field had its own intendance officer who was directly responsible to the intendancy of the local military district, thus allowing the division to move about the supposed field of operations with the minimum of confusion in internal logistic administration.[23]

Rearmament of the Russian imperial army also occupied a significant role in Miliutin's modernization program. The importance of the superiority of the Western Alliance's rifled cannon and other technological advances had not been lost on the observers of the military operations during the Crimean War, and the need to integrate these new technologically advanced weapons into the structure of the Russian army found expression in Miliutin's assertion in the *vsepoddanneishyi doklad* that

> with regard to the contemporary composition of military art artillery technology has received immense importance. Today perfected weapons give a decisive advantage to that army which, in this relation, outstrips the other. We are convinced by this truth from the bitter experience of the last war. Our troops, too late in being equipped with rifled weapons which quickly achieved a superiority over smooth-bore weapons, had to suffer heavy losses and only their determination redeemed the imperfection of their weapons.[24]

Considering technological rearmament of the army as one of the most important tasks to be addressed in the overall program of reorganization, Miliutin turned his attention to the development of domestic industry for manufacturing armaments and munitions. He found these at an extremely low level of production. In the *vsepoddanneishyi doklad* he proposed that in order to bring armament and munitions production up to the level where they would be on a par with those of Western Europe, there should be a purchasing program from foreign manufacturers to meet both goals of timely rearmament and military-industrial self-reliance. In a letter to the state controller responsible for the military industries, Miliutin emphasized his desire to promote domestic military industry with the observation that "Russia is no Egypt, no Papal state, which can limit itself to the purchase of foreign armaments for the whole army. We must build our own factories for the future manufacture of our weapons."[25] All that was left to decide at this point was whether investment would be in state-sponsored as opposed to private enterprises. When this question was brought up for discussion in the Imperial Russian Technical Society in 1873, such prominent members of the scientific community as V. L. Chebyshev and N. V. Kalakutskii formed the majority and favored investment in state-owned enterprises, fearing that the war ministry would otherwise become too dependent on an underdeveloped private industry.[26]

With this recommendation in mind Miliutin attempted to establish control over military-industrial policy through the nascent small arms commission within the main artillery administration of the war ministry and through the allotment of orders to arms factories operating on Russian soil, be they state or private, Russian or foreign-owned.[27] Although this implied a certain amount of rationalization and centralized control over domestic military-industrial development, the arms and munitions procurement process was still rather haphazard. The mining department of the ministry of trade and manufacturing and the ministry of the navy also had stakes in the supervision of the manufacture of artillery within Russia and thus, to a large extent, precluded any effort to establish a centralized agency which could impose a coordinated military-industrial procurement program.[28]

The other important component of the extensive organizational reforms being planned was the need for a highly professional, dedicated officer corps which could assume the greatly increased administrative duties brought about by the reforms. The experience of the Crimean War and the few years of peace after it had revealed the dismally low level of initiative and morale symptomatic of the officer corps as a whole. For Miliutin, the obvious task was to reform the entire military-educational system—from junker and cadet schools to the higher military-educational institutions such as the Nicholas Academy and the engineering and artillery academies.

Although Miliutin was to briefly mention his intentions to reform the military-educational institutions in the *vsepoddanneishyi doklad,* it was not until February 10, 1862, that he was able to devote full attention to the subject in the form of a note to the chief of the military-educational institutions, Grand Duke Mikhail Nikolaevich, entitled "An Opinion on the Military-Educational Institutions." Expressing his dissatisfaction with the end-product of the military-educational system, Miliutin concluded that the system was inadequately training an insufficient number of officers for an exorbitant amount of money.[29] In his opinion the cause of this was the confusion of two entirely separate goals: the need to raise the educational level of the general civilian population and the need to provide higher, more sophisticated military training. "The education of boys and young men should be completed at home and in civilian institutions. Institutions which are strictly military can exist for one purpose only—to provide scientific, specialized education to those young people who feel called to military service."[30] A system which relied on both educational goals rather than unloading the empire's entire educational burden on the military schools was more likely to promote those cadets who showed talent and ability in military affairs and at the same time allow specific military training to be based on an already well-developed sense of creativity and moral strength provided by civilian education. Such was the essence of the problem as Miliutin saw it, one on which he was to base the future structure of the reforms in military education.

Institutional reform of the military-educational system did not actually begin until the special committees for general reform were convened on October 12, 1862. The members of this specific special committee, chaired by Grand Duke Mikhail Nicholaevich and consisting of the war minister, the minister of national enlightenment, and the military-educational department, were all sympathetic to the war minister's view on the subject. The opening statement by the Grand Duke showed his agreement with the war minister:

> To separate the general classes from the specialized, and to reorganize the latter in such a way that young men, preparing for the military profession, are adequately placed in an environment more precisely suited for a military education and actual military service; the general courses to be built on other principles. Finally, to seek methods for the general reduction of military expenditures on military academies and consequently divert the money saved to the Ministry of War and the Ministry of Public Enlightenment.[31]

During the fall the committee reviewed these issues and, when at the end of November 1862 the basic thesis of this work was submitted to Alexander II, implicit in one of the main provisions was the intent to separate the higher military-educational institutions from the cadet and junker schools.[32] This was, in effect, done in January 1863, when Miliutin transferred the Nicholas Academy from its former subordination under the main administration of military-educational institutions to the general quartermaster of the main staff; and later, with the formation of the main staff, it was placed under the jurisdiction of that institution.

This reorganization allowed a greater degree of specialization, although between the cadet schools and the academy this did not pose much of a problem. What was more significant was that now the Nicholas Academy was united under the institution it was to serve, the main staff, thus allowing for more direct communication between the academic and professional responsibilities of the general staff. In what was to be a reversal of the admittance policy established in the post-Crimean War era, Miliutin decided to raise the academic standards of the academy by increasing the number of years of service in the ranks required for admittance, thus cutting the numerical burden of students on the teaching staff.[33] The motivation for this extensive reorganization of the military-educational institutions which Miliutin shared with the newly appointed chief of these institutions, Major-General Isakov, was the recognition of the many deficiencies which had characterized the military-educational system before the Crimean War in conjunction with the sudden growth of revolutionary movements among the cadets at various military schools after the war.

Along with this rearrangement of the academy's main sponsor within the hierarchy of the central military establishment came the beginning of reform within the curriculum to suit its new role. In November 1862 Baumgarten was relieved of his post as the director of the Nicholas Academy to be replaced by General Leont'ev, who immediately took advantage of the new provisional regulations for the academy created while it was under the supervision of the general quartermaster of the main staff, and which gave him more freedom in controlling the day-to-day administration of the academy.[34] Operating under the provisional regulations, Leont'ev proceeded to build up the significance of the academic conferences by making them responsible for all academic and most administrative decisions concerning the academy's functions. The academic conference was to provide a detailed account of its activities in the publication of an annual report, the first of which, for the 1862/63 academic year, showed the remarkable changes which had occurred within the academy as a result of its former instructor's army-wide reforms. It commented on the fact that relatively few of the graduating students actually went into the general staff, as well as criticizing its own curriculum on non-practicality.

In a bid to train these officers exclusively for duty in the general staff, the report proposed to solve the first of these problems internally by limiting admission into the academy to exceptionally talented officers who had already served four years in the field.[35] As far as its function as a distributor of military knowledge throughout the army was concerned, the report stated:

> This last task will certainly be fulfilled [because] the Academy will graduate officers who are fully trained in accordance to their calling: because the duties of the general staff in the Russian army lie not only in the efficient management of the movement of troops, their disposition and activity, but also in the responsibility of instructing the troops. Thus to officers of the general staff lies the responsibility to teach and disseminate literacy amongst the troops; they alone, over other officers, concern themselves with the instruction of military science in the junker schools, cadet corps and finally, in their own Academy.[36]

Throughout 1863 Miliutin supplemented the activities of the academic conference with the formulation of new official regulations for the academy as well as discussing improvements to its curriculum in an attempt to correct the lack of proximity between the curriculum of the academy and the practical needs of the general staff.[37] Special commissions with the academic conference were set up to discuss these separate issues. The first one, for instance, was created to discuss the rights of professors[38] as well as to review the assertion that there was no need for so many instructors when there were fewer students.

The other and more significant special commission discussed all the virtues of adopting the then fashionable Applied School which was greatly influencing the educational philosophy of the French and German war academies.[39] Originally inspired by the military-educational reforms of Scharnhorst during the reconstruction of the Prussian army during the Napoleonic wars, the Applied School or applicatory method after 1854 was regaining its former importance in the German general staff.[40] Instead of simply memorizing the evolution of opposing troop movements from the campaign history of a famous battle, the student was to actively exercise his judgment by suggesting possible scenarios to an operational problem presented in a field exercise or on a map. In this exercise the student presented his knowledge and degree of judgment in a written essay which was to cover the proposed circumstances he faced, the goals he sought to acquire, and how he would accomplish these goals.[41] It was then the role of the instructor to evaluate these essays according to their logical progression and degree of sophistication. The ideal of the applicatory method was to reemphasize informed contemplation in the curriculum and an awareness of the circumstantial nature of warfare, an ideal which von Clausewitz had originally attributed to Scharnhorst: "He taught that part of the art of war which until then had scarcely been dealt with on the lecture platform and in books—war as it actually is."[42]

The members of the special commission and the academic conference had become familiar with these new techniques through N. N. Obruchev, professor of military statistics and geography, who had observed them at the *École Supérieure Militaire* while on study-leave at the French general staff in the early 1860s. In his original report to the academic conference, Obruchev observed that not all of this system was suitable and that it would have to be carefully adapted to the higher level of theoretical training extant in the Russian military-educational system. Returning to Paris in mid-August 1864 to become more familiar with the French Applied School, he paid a visit to the aging founder of the Nicholas Academy, Jomini, to discuss the changes in the academy's organization and curriculum. Jomini at this time was interested in the techniques of the Applied School, despite the fact that it was originally developed to counteract static, normative theories, and among a series of contentious observations suggested that a supplementary year be devoted to "applicational exercises"—a suggestion which was eventually realized by the academy in 1868.[43]

By October 1864 the deliberations of these special commissions at the academy came to an end and their findings were summarized by Men'kov:

> The extremely vast composition of a few of the theoretical courses at the Academy were excessive and onerous obstacles to the students, and contributed to the loss in fundamental instruction in these subjects as well as presenting an obstacle in the Academy's development of many other practical subjects which would have been very helpful to officers of the general staff.[44]

Eventually the report from these commissions combined with the annual report of the academic conference provided the material for a special meeting on the reexamination of courses, which was convened by Miliutin in January 1865.[45] Having brought together all the members of both deliberating bodies, Isakov,

Men'kov, Bogdanovich, and others, Miliutin personally began the proceedings with an observation:

> The officer of the General Staff recognized his duty to lead the battalion at his disposal but saw those battalions only as tactical units; the internal organization of the troops, the way of life and needs of the soldiers were hardly familiar to him; in the actual command of the troops he almost never took part. A similar situation cannot continue with regard to reform taken from previous times. From officers of the General Staff the government demands fundamental familiarity with every internal mechanism in the command of troops.[46]

He added that the creation of the military district system now required a familiarity with all mechanisms of the domestic administration of troops in addition to all former responsibilities, and that such training would best be provided by the Applied School format under the subordination of affiliated departments in the hope that such a curriculum would train highly capable officers, not more military theoreticians.[47] The remainder of the meetings discussed specific aspects of the curriculum, such as the teaching of military art and military history and how the academic conference was to carry out its obligations. "Thoughts upon Re-Examining the Courses and Academic Exercises at the Nicholas Academy of the General Staff" was published as a result on February 25, 1865.

On the same day, the war ministry issued its own report expressing its approval of the findings of the special meeting and praising the accomplishments of the academic conference in particular for its ability to efficiently administer the affairs of the academy.[48] Miliutin's three original goals for the reform of the academy's curriculum had remained intact. If the academy was to prepare officers for duty in the general staff, then it faced a tremendous task in preparing these officers to assume the administration of the entire army, a task best met by the highest degree of practical training. The ministry's conclusions on how to instill such a practical direction in the curriculum presented a fundamental challenge to the study of strategy and military history and criticized the enormous growth which had taken place in both these subjects under the teaching of Bogdanovich:

> Too much development of historical data consumes much time to the loss of practical exercises. Criticism which is not based on detailed analysis of collected facts can promote indisputable verdicts. Finally, excessive dogmatism in the military sciences can lead to one-sidedness and abstract theories completely divorced from direct practical goals of a specialized military education.[49]

The alternative pedagogical method which the report suggested would provide a more practical direction to the instruction of these courses. The methods of the Applied School were combined with the critico-historical method developed by Lieutenant G. A. Leer. A former student of Bogdanovich, Leer explained the practicalities of his proportional balance between descriptive history, historical analysis, and logical elements of theory in an article published in *Voennyi sbornik* in 1863.[50] This alternative method divided the teaching of military history into two types of analysis. The history of military art focused on the continually developing trends throughout the ages, and was closely coordinated with the teaching of strategy (in fact, it made up the introductory part to the course on strategy). The second part distinguished military geography and statistics. The functioning of the eternally developing trends in military art were to be analyzed in the context of a single campaign to illustrate their interdependence, so eloquently explained by Leer:

> The course on military history as a collection of critical reviews of completely separated facts is, of course, unthinkable. The connection [between them] is possible in the following two forms: A.) a review of critically collected facts in a sequence in which questions on the combination of elements which stand, for the most part, on their own (for instance, the consideration of plans, i.e., review of the subject of movement, bases and operational lines, the further concentration of forces, the combination of marches, manoeuvres, and battle) and B.) the critical review of one or a few of the most significant campaigns in full order (for instance, the campaigns of 1796 . . .). Both these methods must be used in totality to establish a synthesis of the parts (combination of elements) of strategy in the critico-historical form.[51]

Actually, it was the combination of both types of analysis which was to make up the course on strategy and such a pursuit placed emphasis on the importance of detailed information, implicitly recognizing the need for more archival development.[52] The report went on to discuss the other courses which made up the main curriculum, those of military administration and military statistics.

A review of the content of these two courses was conducted in order to eliminate any overlap.[53] The purpose of military administration was to acquaint the officers with the military and government institutions which had administrative relations with the functions of the general staff. The second part of the course was on the direct administration of the troops within the military district. The practical exercises in conjunction with this aspect of the course concerned clerical work and other paper-work exercises pertaining to the bureaucracy of the military districts.[54]

Conversely, the first section of the course on military statistics, involving the compilation of statistical information concerning the Russian imperial army, strongly resembled those exercises covered by military administration and, at the time of the report's publication, this redundancy had not been resolved. However, military statistics had sole jurisdiction over the compilation of statistics on foreign military establishments (as it had always done) and it also was to be accompanied by practical exercises involving the collection of military-statistical information and the written resolution of various questions concerned with a given region. In addition, the technicalities of the topographical exercises concerning geodesy and cartography were also discussed, as well as the auxiliary subjects of artillery, engineering, fortifications, political history, Russian, and foreign languages.[55]

With the major curricular issues resolved, the special meeting turned its attention to the administrative changes necessary in the academy. The first question was whether four years of military service as prerequisite to the academy, a policy which most of the members supported, was appropriate. As to enlarging the curriculum to meet the additional demands of training, the special meeting concurred that starting the academic year earlier, in conjunction with expanding the number of lectures per day to four, starting at 9:15 A.M. and going until 3 P.M., would satisfy these demands.[56]

The findings of this special meeting were then used the following year to compile the official thesis for the academy, which was begun in January 1866 and completed in March 1868.[57] For the most part the thesis followed the recommendations of the report but made several changes. Along with the adoption of a supplemental course, the consolidation of tactics, strategy, and the two military histories into one faculty of military art was the most obvious alteration in the curriculum and was due to the large amount of redundancy between these subjects.[58] It also restricted policy on hiring instructors, noting that over the duration of the previous four or five academic years very few candidates for teaching positions at the Nicholas Academy had defended a dissertation.[59] It also clarified the functions of the academic conference and the requirements for membership in that body.[60]

The effect of this series of army-wide reforms was to raise the general staff to a prominent position in the Russian army. In addressing these increasing professional responsibilities, the reform of the Nicholas Academy not only embraced the original program which Jomini had presented to Nicholas I in 1826, a vocational and academic program to encourage "love of study," but extended this mission to the promotion of a new professional ethic based on practical competence for dissemination to the rest of the army and ultimately to Russian society at large.[61] Although many aspects of the Russian army were to change over the subsequent decades, the academy's educational program was to remain relatively constant until the validity of its curriculum was brought into question in light of Russia's defeat by the Japanese nearly forty years later.

By the early 1870s the mansion on the *Quai Anglais* which had housed the academy had grown into a magnificent structure suitable to the increased stature of the general staff. The offices of the faculty were located on a circular ground floor. A grand staircase led to the second floor where two lecture amphitheaters for the first- and second-year students were located, as well as the library, laboratories, and examination rooms. On the walls of the first-year amphitheater were portraits of Peter the Great and prominent Russian generals, with a portrait of Napoleon in the place of honor. In the second-year amphitheater resided a single painting of Napoleon accompanied by his entourage—an indication of the great reverence held for the man who had become the symbol of genius in military art.[62]

As prescribed in the thesis for the academy, the academic conference in consultation with the war minister was responsible for proctoring the admission of new students and maintaining rigorous selection standards for the hiring of new faculty as well as supervising the content of the academic curriculum. Ideally, officers who, after four years of service, had achieved the rank of captain, second rank in the Guards, captain in either of the technical arms, or major in the infantry or cavalry[63] were eligible to sit a preliminary examination administered by the chief of staff at the headquarters of the prospective applicant's military district. Those who passed this exam were allowed to present their credentials to the director of the academy for admittance to a second examination at the academy in the first week in August of every academic year. A second oral examination, presided over by an examination commission appointed by the director, tested each applicant in the following subjects: elementary mathematics (arithmetic, algebra, geometry, and plane trigonometry); political history of European nations up to 1815; physical and political geography of the five continents; details about the Russian Empire; army rules and regulations and the tactical regulations for the various arms; the basics of artillery ballistics and fortification; and the elements of military administration. Examinations in the Russian language and one foreign language—either German, French, or English—consisted of a composition written on a given topic and the translation of an article on a military topic, respectively. On a scale from 1 to 12 the applicant could only pass with a grade above 8 in Russian and an above-8 average in the other topics with no grade lower than a 6.

All in all the admissions examination tested the general knowledge of the applicant and, in comparison to St. Cyr or the Berlin academy, did not insist on more detailed knowledge of military affairs.[64] Although the majority of students were better prepared for the admissions examination than their predecessors before the era of reform, they tended to be weaker in foreign languages and geography. In the decade after the reform, the academy was successful in limiting the admission of officers to an average of thirty-five new students, but after the Russo-Turkish War figures shot up to 150, a peacetime educational program which far exceeded the admission levels during the Crimean War.

For those capable few who passed the exam to obtain a place at the academy, the first-year curriculum was split between principal and secondary courses. The principal courses, which were taught exclusively by the permanent faculty at the academy, were the history of military art, history of famous campaigns, tactics, and strategy (all of which, despite the ruling of the thesis, continued to defy adequate delimitation); military administration, which later became the dominion of mobilization and railroad schedules; and military statistics, which remained the least favorite course among the students. In an attempt to introduce the students to contemporary military art the academic conference, true to the thesis, cut out material on ancient military history and ignored the campaigns of the eighteenth century to concentrate on the war with Poland in 1830/31, the suppression of the Hungarian rebels in 1849, the Crimean, Italian, American Civil, and Austro-Prussian Wars, and the Franco-Prussian War of 1870/71.

For the tactical instruction of both first- and second-year students, groups of six or seven would be assigned to a general staff officer from the St. Petersburg military district. This allowed a greater amount of personal attention to each student over a long period of time which produced good results and proved to be an excellent way of spotting new instructors for the regular courses. The secondary courses were taught by preceptors, often professors from the University of St. Petersburg, who taught general history, philosophy, psychology, geography, topography, and Russian. As far as the teaching of foreign languages was concerned, the majority of the students studied German throughout the last quarter of the nineteenth century. Even at the high point of Franco-Russian relations in 1896/97, eighty students were learning German compared to forty-nine who were learning French, much to the chagrin of a French officer who visited the academy at that time. In the summer term the students went on mock staff rides and topography field trips around St. Petersburg, practices which had been brought back from France by Obruchev along with the applicatory method.

Upon successfully completing the first-year course (or skipping first-year course work by examination), the students entered into a second year which stressed more involved instruction in strategy, the tactics of larger bodies of troops, administrative exploitation of railroads for military uses, as well as a course on using the telegraph for communication (introduced into the curriculum after the Russo-Turkish War of 1877/78). During the winter months the students in their study-groups would continue to do tactical map exercises beginning with a two-week home project on the scenario of an army corps in combat, and another week-long project on its march maneuver. Later they conducted "flying problems" (*letuchaia zadacha*), tactical map problems to be solved within three hours. In addition to this normal course load were obligatory extracurricular presentations on military-historical subjects given by the professors and occasional lecturers from outside the academy. Students who performed satisfactorily in the examination at the end of the second year, but who did not excel enough to obtain a place in the third-year supplementary course, reentered the service with a year's pay and the right to wear the insignia (*akademicheskii znak'*) of the academy as well as the chance to be more quickly promoted to lieutenant-colonel, the first rank of superior officer. Even if the student had not completed the general staff supplementary course, his credentials were still regarded in the Russian army as an indication of a high degree of professional ability. As Obruchev was to observe, during the Russo-Turkish War a general officer who gave an order to a line officer wearing the Nicholas Academy insignia had more confidence that his order would be successfully executed.[65]

For those students who excelled in the first two years there was the final year made up of the supplementary course with enrollment strictly limited by the minister of war. The purpose of this course was to cap off the applicatory training of the students who would be going on to service in the general staff and to introduce them to academic research. The first assignment was a thesis or logical account of a specific military campaign using original sources. The second was an oral presentation on a theme from military art to be given at the academic conference. The third, and by far the most difficult, was a strategic and tactical applied problem divided up into three parts, designed to simulate administrative and operational preparations for a war. Given command of a hypothetical army corps and a statement of the general political and military-operational situation, the student was to first perform a military geostatistical evaluation of the theater of war and then begin to manipulate the strategical and tactical operations of the army corps. The third part was to develop these same strategic and tactical operations from an administrative point of view. During the 1869/70 academic year the director of the Nicholas Academy, General Leont'ev, instructed the members of the academic conference to draw up regulations for the conduct of these war games suitable for instruction to the officer-students, the result being the publication in 1873 of Colonel Kazanskii's *Guide to War Games* followed shortly by the work of Puzyrevskii, assisting Stankevich and Gudim-Levkevich in the practical exercises for tactics. Puzyrevskii published an article in *Ruskii invalid* entitled "War Games and a Guide to Their Conduct" in 1875.[66] Throughout the six-month period of the supplementary course the students were constantly evaluated by members of the academic conference, and their future placement in the general staff was determined by their performance during this time.

Along with this increased emphasis on applicable knowledge and love of study was a new-found respect for technological inventions which promised to transform the nature of warfare. In 1866 the Imperial Russian Technological Society was established by several leading Russian scientists and military professors to address the excess of scientific specialization in the military and civilian scientific community.[67] Strongly supported by Alexander II from its inception, the society immediately grew to six regional chapters; the Siberian chapter had 1,250 subscribing members. Each chapter consisted of four sections: (1) applied chemistry and metallurgy, (2) mechanical technology, (3) civil engineering and architecture, and (4) naval engineering, artillery, and arms manufacture.

Free to discuss any technical question, even those under review in government circles or in the higher military administration, the chapter committee would select from its members technical questions for discussion in public lectures, the proceedings of which were published in the society's bimonthly magazine. On issues concerning more specialized military matters, the society would hold winter conferences at the various military academies for students and officers of all ranks. The conferences held at the Nicholas Academy (restricted to general staff officers) were the most numerous and were occasionally attended by army commanders, members of the royal family, and the emperor himself.

The changing nature of warfare apparent during the Crimean War was not lost on many of the instructors at the academy, as seen in the work of Astaf'ev and others. Yet its implications on the appropriate training of officers was to draw attention only in the 1860s and 1870s. The center of attention was to focus on the teaching of military history, a course taught by Bogdanovich from the early 1840s; a reevaluation of its contents was well under way by the time he retired in 1863.[68]

Although a symbol of tradition at the academy, Bogdanovich was certainly not insensitive to the widespread changes which were occurring in every aspect of military affairs. Upon retiring from the faculty of strategy, military history, and military literature, he devoted his attention to the growing importance of developing military historical archives under the auspices of the newly formed military-academic committee of the general staff. Created by order of the war minister, the military-academic committee was involved in promoting military-historical research by promoting the systematization of the military-academic archives in St. Petersburg and at the department of general archives of the general staff in Moscow as well as sponsoring a concourse for military-historical monographs and the translation/publication of many foreign military historical works. Although Bogdanovich had advocated the use of archival sources throughout his career, his later works benefited a great deal from his systematization of the archives while at the military-academic committee, and he was responsible for introducing Russian military historiography to the growing sophisticated use and management of available information.[69]

The reform of the course of military history itself was left to Bogdanovich's former students, Leer, Dragomirov, Berens, and Stankevich. These men were to play a significant role in the review of the curriculum at the special meeting and were largely responsible for differentiating the study of continuous elements in the conduct of warfare from the study of the interrelationships between these elements within the context of a particular campaign.[70] With such a differentiation established, Berens' lectures on strategy, drawing mostly from his wide reading of military literature, were to make up the new text on the history of military art.[71] Stankevich was to take another approach, one originally suggested by Astaf'ev's work: the evolution of trends in military art from the study of the composition and organization of troops, military institutions, means of armed conflict, systems of supply, and the organization of the rear services.[72]

Occasionally the course work was supplemented by lectures from outside the faculty. For instance, the professor of military statistics, Colonel Maksheev, described the characteristic conduct of warfare on plains regions and gave a short presentation on the military influence of Russia in East Asia.[73] Beginning in 1867 a reevaluation of the course on military statistics was conducted by Maksheev with the aid of Obruchev, who had recently been appointed head of the military-scientific committee of the general staff. Drawing from his sabbatical of six months to the Turkestan military district, Maksheev introduced many new courses more oriented toward Asia, such as a course on military-political relations with Asia which analyzed geographical, historical, military-historical (including the exploits of great Asiatic commanders), ethnography, statistics, and economic and political features of China, Central Asia, Afghanistan, and Persia.[74] The other courses included historical-geographical studies of Russian and European (especially British) colonial developments throughout Asia.[75] In the realm of tactics, Dragomirov was to make a substantial contribution to this new faculty with his history of the Austro-Prussian War of 1866 as well as his systematization of military pedagogy in the training of troops.

However, the most influential contribution to the discussion of military art, and in particular that of strategic theory, was made by Leer. Beginning his career as a field engineer in 1848 and entering the Nicholas Academy in 1852, Leer had been an avid student of military history under Astaf'ev and Bogdanovich. Although his performance at the academy was marred by family problems and a lack of talent in sketching, Leer managed to obtain an adjunct-professorship in tactics in 1858 on the basis of his talent as an instructor and writer on military-historical themes.[76] In 1860 the academic conference assigned him the task of forming a new course on military history to facilitate the students' critical research of questions pertaining to elements of military art in a rapidly growing field of knowledge.[77] Motivated by this assignment, his work alongside Berens and Stankevich, his course on military art at the engineering academy, and the report of the special meeting in 1865, Leer gradually developed his ideas for such a course, represented throughout the period 1858–1865 by many of his articles published in *Voennyi sbornik*. By the end of the 1860s Leer felt he could synthesize the many different facets of speculative and empirical analysis to more adequately introduce the many concepts of military art in a straightforward manner to the students at the engineering academy.

The result was the publication of *Attempt at a Critico-Historical Approach: Towards a Positive Strategy* (*Opyt kritiko-istoricheskogo issledovaniia zakonov iskusstva vedeniia voiny* [*polozhutel'naia strategiia*]) in October 1869. This work represented a significant jump in the philosophical sophistication of Russian military-theoretical research. As evident in the title, Leer's intention was to combine the philosophical legacy of von Clausewitz, the teaching of his former professor, Bogdanovich, and the idealism of scientific positivism which was popular in educated European circles at that time. Explicitly stating the fundamentally positivist assumption that "facts are the living embodiment of important [eternal] ideas" and that action was a direct result of thought, Leer introduced his study with a comment on the evolutionary development of a still primitive branch of knowledge:

> At this moment science is in the "positive" period of its existence having gone through, depending on the quality and quantity of analytical work, the following periods: the metaphysical (period of fabrication and hypothesis), the scholastic (period of undeniable laws), and the dialectical (period of denial, casting doubt on all previous analysis; a period of discussion and inaction).[78]

In his opinion such a positive development of a military science could be obtained only by a careful combination of historical description and speculative analysis mutually complementing each other—in other words, the strict application of the criticohistorical method he had first promoted in his article in *Voennyi sbornik* six years previously.[79]

With this flourish of philosophical plumage, Leer launched into the discussion of theory by coming to grips with his own philosophical assumptions about objective reality. He devoted the first section to a discussion of the nature of war and its social function in the life of a nation. On a fundamental level Leer combined the saliency of ideas with von Clausewitz's preoccupation with dialectical struggle in social phenomena to reach a conclusion which had Darwinist overtones. War was a highly natural phenomenon in the relations between peoples and was actually a positive and progressive dynamic in the cultural growth of a nation.[80] This was evident in comparing the social structure of a nation before and after a war. The crisis of war would most certainly point out national weaknesses and stimulate social reform regardless of whether the nation had won or lost. Leer most probably had in mind the example of the Russian Empire's defeat by the Entente powers during the Crimean War and the many enormous socioeconomic and military reforms which had occurred in the years directly after that war. Other historical examples could also be enlisted to show war as the most powerful and efficient "civilizer of nations." For example, Alexander the Great through his military conquests was able to introduce Greek culture into Asia within a relatively short time. A more oblique aspect of this philosophical hybrid of Kantian ideal and Hegelian dialectic was that ultimately the development of military art and a nation's cultural sophistication were directly interrelated—a concept which was later to result in misinterpretation and a storm of controversy.

Proceeding from the consideration of strategy as a component in the science of human behavior to its more operational concerns, Leer addressed the central mystery of military art which Tolstoy had recently illustrated in his novel *War and Peace:* "History shows that numerically smaller forces often win against superior numbers." Whereas Tolstoy was to account for this paradox by focusing on the "spirit of the army," Leer was to take the opposite and more traditional view that it was the result of superior genius on the part of the individual commander. In this respect Leer was greatly influenced by von Clausewitz in his discussion of genius in the execution of military art, and was to compare it with that of other types of artistic expression in the delightful observation that "military art is more difficult and more complex than others because in the other arts the elements are passive, inanimate objects whereas in military art the elements are people as the main weapon of war which add many different dimensions to the art."[81] In addition, genius was to maintain its original description in the concept of resourcefulness (*nakhodchivost'*) which was made up of the traditionally understood elements of intellect and will. Giving examples from great military writers, Leer placed emphasis on the latter, agreeing with Napoleon's maxim that war was three-quarters will and one-quarter intellect.[82]

The other shared concept was the personal capability to judge by eye (*coup d'oeil*), or as Leer was to call it, military diagnostics, a concept traditionally known in Russian military speech as *glazomer.*[83] He felt that this spontaneous combination of resourcefulness with an ability to judge the various interactions between circumstantial phenomena on the field of battle defined a true military genius and had only been accomplished by a few military commanders throughout the history of warfare. However, in reaction to the growing complexity of conducting warfare so apparent in the experience of the Crimean War and other European and North American wars during the 1860s Leer, like von Clausewitz, was trying to supplement what had hitherto been subjective personality traits with a critical analysis and systematic code of empirical trends in military art.[84] Military art was a huge theoretical scale or, as Napoleon put it, "the axis for constructing [nonlinear] geometrical functions," consisting of a few positive laws serving as a basis for military art but made conditional in their application by the limitless potential circumstances of the battlefield.[85]

The theoretical balance responsible for accurately determining the correlation of ends and means (both subjective and material) was to synthesize the whole range of military affairs, from its most abstract to its most mundane. In agreeing with von Clausewitz that war was state policy conducted by more extreme means, Leer stressed this correlation of ends and means in the observation that "the military goals of a nation must be placed in strict accordance with its [actual] strengths and means, in general with the degree of its national richness," concluding that overextension of military goals could lead to the deterioration of domestic and foreign political relations. Principles of a subjective nature garnered from criticohistorical analysis, such as the old Suvorovian principle of firearms as preparatory weapons and the cold steel of the bayonet or sabre as shock weapons, provided the subjective correlation between the goals set by the resourceful genius of the commander and the moral capabilities of his troops. Finally, the Napoleonic principle of concentrating one's forces at the decisive location and time provided the operational variant of this need to correlate material and moral resources to the dictates of circumstance. These three variants systematically utilized in both their speculative sense to determine the general nature of future events and their more empirical analysis of the wealth of past experience for insight into specific operational forms of military art were to form the basis of a wider theory of military art (strategy) or positive military science.[86]

The practice of the criticohistorical method could be further characterized by a differentiation of subjective and material components, a typology which Leer claimed was first introduced in military-theoretical writing by Lloyd and adopted into Russian military-theoretical thought by P. A. Yazykov. Leer emphasized the importance of the subjective component by claiming that the greatest extent of the parameters for this element was the spiritual strength of the whole nation playing the largest role in the conduct of war, an orientation which suited his fundamental assumptions about the social role of warfare and the art of the military commander. As usual, he cited Napoleon's maxim that three-quarters of military success depended on the spiritual qualities of an army. He even suggested a new branch of the military sciences not possible before (military-psychological studies), which he felt would undergo significant development in the years to come.

More conducive to a positive science were the material elements of military art which could be broken down into many specific typologies. The first one focused on the human material which comprised the most important weapon of the army. This element formed the heart of tactics, the most advantageous use of troops and their organization in accordance with known goals. Elements of terrain were also important, especially in studying the theater of military operations for the formulation of a war plan by taking into account aspects of military geography, topography, and fortifications. Civilian and military administrations shared a concern with the interaction of state and military organizational structures on the efficient supply, formation, and mobilization of troops (a subject which interested Leer's colleague Stankevich, and which later became the heart of criticism against Leer's methodology). Military-statistical analysis was concerned exclusively with the hardware component of military power of both the Russian Empire and her immediate neighbors at any given moment.

The last and by no means insignificant element was that of chance, the great god of war. Leer made a distinction between internal and external aspects of circumstance, noting that the internal aspect was specific to the capriciousness of humans differentiated from the capricious nature of the environment (weather, inaccurate information, etc.) and cited many statements from famous military writers to illustrate the significance of this phenomenon.[87] Warfare demanded the utmost effort from the individual commander in synthesizing all these elements, giving definition to the highest form of military art: "The influence of these (elements) upon each other and their combination in dependence on time and space for the successful attainment of war goals makes up the synthesis of all military affairs, Strategy."[88]

Having elaborated on the essence of military art and given a definition of the concept of strategy, Leer turned next to the Clausewitzian dilemma: "Has Strategy existed long as an art and a science?" On this topic he held resolutely to von Clausewitz's original argument: strategy as an art had existed as long as warfare. To support this opinion he cited the example of Alexander the Great indulging in one of the eternal practices of strategy, the preparation of the theater of military operations. However, it only existed in the actions of select personalities who had an instinctive, natural grasp of the practice of strategy; in other words, strategy was exclusively an art until recent history because it only found haphazard expression through the genius of a particular commander.[89]

Not until Lloyd's attempt to apply rational analysis to the phenomenon of warfare in the 1780s and 1790s was there any move from subjective, instinctive art to a more conscious codification of military principles. Leer had a high regard for Lloyd's pioneering work, granting him credit for originating the link between military operations and political considerations, the concept of the operational line, and a sophisticated view of tactics similar to those of a contemporary military genius, Suvorov.[90] He also noted that Lloyd had been a truly sophisticated military thinker and had often been misinterpreted by successive military theorists. As an example of Lloyd's flexibility in historical-theoretical analysis, Leer pointed out that he had acknowledged deviations from his theory concerning the use of operational lines common to large European armies, noting the Tartar method of requisitioning rather than relying on complex lines of supply and communication.[91] Leer then turned to a review of subsequent military theorists who tried to address the creation of a military science, assessing and often criticizing the limitation of each and developing the different attitudes toward the concepts of offense and defense based on the paradox between the need for decisiveness and the need for safety. The extent of his familiarity with both Western and Russian works was substantial and indicated the emphasis placed on abstract theoretical literature in the Russian higher military-educational system at that time.[92] After reviewing individual attempts at analysis and the meagre attention to synthesis displayed by Western military theory, Leer came to very interesting conclusion:

> The distinguishing of features in strategy is impossible. The final task of theory must consist not of the sharp distinguishing of its features but only to show how it incessantly flows throughout each military question, from one to another. Finally, if we undertake and subject some facts to critical review and from that shade in the strategic aspect of affairs then we get strategy. If, in selecting these facts we shade in the tactical side then we get tactics.[93]

For Leer, bad theory consisted of overanalysis which separated elements from their interrelation with other elements and generated obscure definitions and classifications.[94] Leer gave an example of a possible format by which one might go about examining the links tying together different aspects of strategic questions.[95] His work immediately drew the attention of critics who were either not impressed with his claim to a positive strategy, finding him overly speculative, or who were dissatisfied with his inability to bridge the gap between theory and its practical application.[96]

Although this first work was to provide a lucid and thought-provoking argument in favor of a criticohistorical method as foundation for a positive military science, the combination of these two philosophical traditions was to create a host of problems which were to increasingly bedevil Leer's theoretical development in the years to come. The first of these was the incompatibility of a scientific positivism which sought certainty in prediction with a dialectical speculation which could only strive for a degree of probability. The other was the misconception that civilization and military art enjoyed a logical and parallel development, an assumption which eventually promoted incremental, quantitative changes in theory and practice which were to deny more substantial and qualitative realities.

No less a contributor to the development of a theory of military art to match the new professional responsibilities of the general staff was M. I. Dragomirov in the

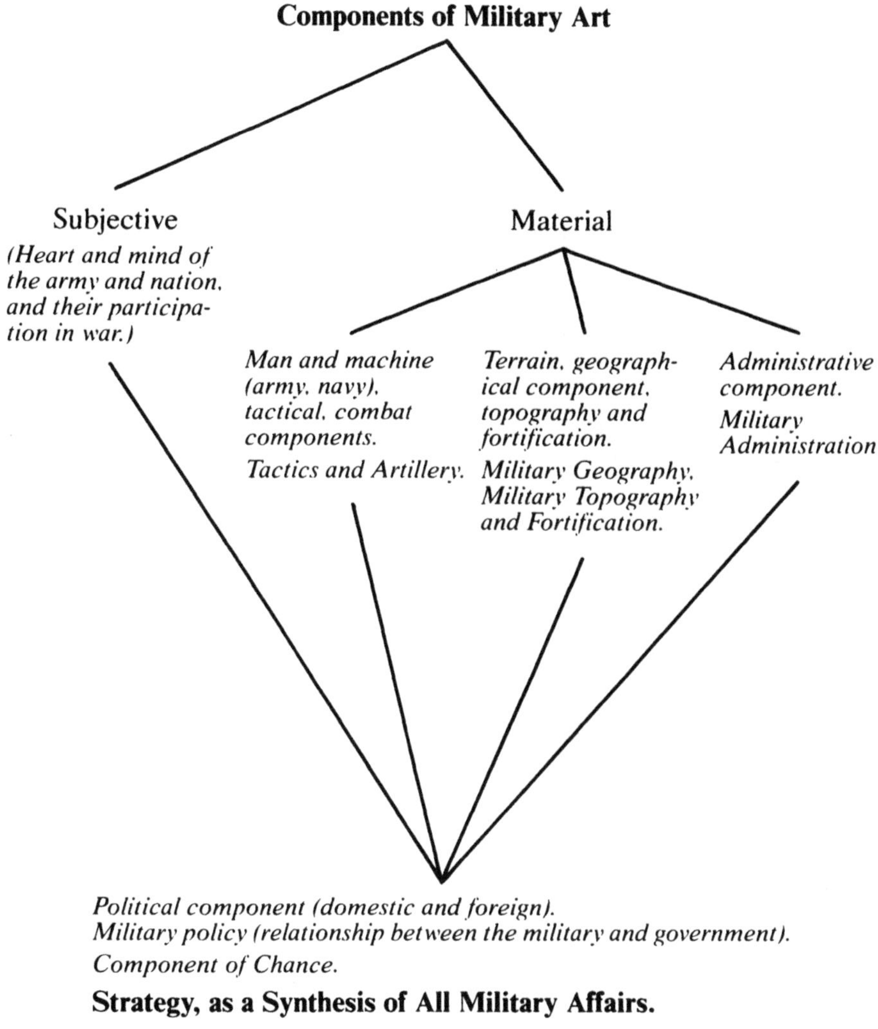

Source: G. A. Leer. *Opyt krytico-istoricheskogo issledovaniia zakonov iskusstva vedeniia voiny.* St. Petersburg, 1869.

field of tactics. Entering the academy's first-year course at the height of the Crimean War, Dragomirov had been spared no antipathy in the lectures of Astaf'ev and Goremykin for the backwardness of Russian infantry tactics. Upon completing his course with honors, he became one of the first of the academy's students to take advantage of foreign study, spending a year abroad studying the French military curriculum at St. Cyr and serving as an observer of French and British training methods.[97] Appointed as professor of tactics at the academy in 1860, he emphasized the new applicatory method of training infantry tactics which was to become the combat doctrine of the Russian army well into the twentieth century.[98]

Dragomirov's attitude toward the nature of warfare was essentially a negative reaction to the ideals of the Frederickian formalism which had persisted in the Russian army up to the Crimean War. Contrary to the mentality of the parade-ground and strict adherence to arbitrary and often completely absurd regulations, he saw the realities of warfare in a larger context—a multitude of material and moral forces locked into a deadly struggle which defied logical prediction and presented a severe challenge to comprehension.[99]

In the past the illusory nature of this challenge had been met with static rational prescriptions or philosophical defeatism which assumed that the dynamics of the battlefield was incomprehensible to the human mind and that the only way to cope with it was to deny the genius of the commander and concentrate on a reductionist ideal of interpreting infinitely small details of social interaction to attain laws of historical development. Like von Clausewitz, Dragomirov felt that a proper theory of military art would have to accept an attitude toward history as a "science of incomplete conclusions" formed from only partial knowledge of social interaction. These conclusions, in the form of hypotheses, would provide inspiration to pursue a particular course of action and should never lapse into models for imitation, an attitude which found similarity in other, less complex fields of knowledge:

> This is the method of all the experimental sciences beginning with applied mathematics, the most precise of them all. Take for example topography. The errors which result from the inherent imperfections of the [measuring] instruments themselves and the inaccuracies incurred in their use do not permit exact mathematical maps. Is this a reason to say that these plans have no value? They are imperfect, but nevertheless they are not false. It is true that the idea which they give of the terrain is only approximate, but it is not at all a false idea.[100]

In Dragomirov's opinion the military professional could not afford the luxury of indulging in either prescriptive or defeatist theories. The reality of war demanded the development of a theory of military art (he could never bring himself to call it a science) which shunned reductionism and recognized its relative impotence at the hands of hazard.[101] Commenting on Prince Andre's vacillation between the Frederickian theories of the German strategists and disbelief in any theory whatsoever in Tolstoy's novel *War and Peace,* Dragomirov concluded that humanity could only ask for a compromise: "A little is better than nothing at all."[102]

In this light theory was to play an unusual role in the hands of the military professional in that it was to override the traditional opposition between material and moral factors to approach a calculus of confidence, moral strength "augmented or diminished with the quality of the armament and the troop formations from the point of view of combat, with the number of soldiers opposing the enemy, the confidence in the commander's talents, etc."[103] It was this moral-material composite which was to provide the departure point toward comprehension and eventual manipulation in the context of a reality dominated by hazard. The key to this composite theory was the moral relationship between the commander and his troops. In his review of Tolstoy's *War and Peace,* which was translated and published in France in the mid-1890s, Dragomirov indicated his attitudes toward the role of the commander in his praise of Tolstoy's portraits of Bagration and Kutuzov, noting that they possessed that quality of experience so essential in the commander, the *coup d'oeil* or ability to follow the unfolding of events and effectively convey their recognition of advantageous dispositions (technique) to the troops. This in turn promoted the troops' confidence in the leadership of the commander and heightened their resolve to overcome the instinct for self-conservation and make sacrifices in the face of danger.

The other aspect of the theory of military art was the relationship between a soldier and his comrades-in-arms when they had entered into the actual fray of the battlefield, thus temporarily severing contact with their commander. This was to be dominated by the equally powerful concept of assistance, the spirit of self-sacrifice for the welfare of a comrade-in-arms, and, in a more abstract sense, for the welfare of the nation. Thus in a way the army resembled a beehive where queen and worker bees cooperated in a relationship of mutual support, or an orchestra where the conductor provided essential but intangible guidance to the communal, artistic expression of the musicians. In the context of battlefield hazard, this composite theory was to function much along the same lines as the law of the conservation of energy in the natural sciences:

> There is a truth which is as applicable in the moral world as it is in the material, that is that if someone causes the least change in the composition of something [by this Dragomirov meant the "physiology of a collective organism"—an army], the result is an inevitable change in the size and direction [of that thing].[104]

In this way Dragomirov was to promote increased consideration for the spiritual training of the officer and soldier which was, strangely enough, influenced by the same regard for scientific positivism as Leer's theory of strategy.

As a professor of tactics at the Nicholas Academy and member of the committee on the structure and training of forces in the early and mid-1860s, Dragomirov was inspired by the Suvorovian maxim "to teach only that which is necessary in combat," and sought to provide subjective indoctrination and tactical vocational training conducted through example in field exercises.[105] The essence of this tactical doctrine lay in the correct balance of self-preservation (the domain of the intellect) and self-sacrifice (domain of the will) symbolized on the battlefield by firepower and the cold steel of sabre or bayonet, respectively.[106] Responsible for the formulation of the *Regulations for Infantry Combat Service,* published in 1866, Dragomirov institutionalized this balance in the following battalion attack formation: initial approach to the enemy by skirmishers advancing by leaps and bounds (fifty, one hundred paces at a time) followed by line companies in closed formation. After skirmishers had approached close to the enemy and suppressed rifle-fire, they moved off to the sides and left the line companies to conduct the massed bayonet attack.[107] Ideally this balance between dispersion and concentration of attacking troops was to be a function of the greater interaction of all the components of military affairs represented in his composite theory: "One doesn't need to know in general if concentration is better than dispersion, but only *when* it would be convenient to have recourse to the former rather than the latter."[108] It was the purpose of exercises and maneuvers to further evaluate the proper balance between these two.[109]

Like Leer, Dragomirov's initial work in tactics had its critics. Astaf'ev's original advocation of dispersion and importance of the skirmish line as the main formation for the attack in an era dominated by rifled weapons was to haunt his theory up to the beginning of the Russo-Turkish War in 1877/78. Inspired by the experience of the Franco-Prussian War in 1870/71, a General Zeddeler, possibly a relation of the first vice-director of the Nicholas Academy, gave public lectures and published an article in *Voennyi sbornik* entitled, "The Influence of Breach-Loading Weapons on Infantry Combat Training" in 1872 which was to reach the conclusion that "individual fire [had become] the chief form of combat fire," refuting Dragomirov's holistically balanced tactical formations. However, not being directly connected either to the military-educational community or a member of any military-scientific commission, Zeddeler had no impact on Dragomirov's tactical doctrine at this time.[110]

Within a few years after the publication of his gigantic *Attempt at a Critico-Historical Method,* Leer dominated the development of theory in military art without any serious opposition.[111] The appointment of his colleague Dragomirov to the post of chief of staff of the Kiev military district in 1869 and then the command of the 14th Infantry Division in 1873 left the teaching of tactics to Leer.

Subsequent military events in the Franco-Prussian War in 1870/71 were to provide exciting new material for Leer to demonstrate the full prowess of his criticohistorical analysis. In 1871 he published a campaign history of the Franco-Prussian War which was to be used as the main text for the section on the history of military art in his lecture on strategy at the academy.[112] General Moltke's success over the previously highly regarded French army was to prove a major new inspiration for Leer, who analyzed Moltke's style of strategic leadership by comparing it with that of Napoleon Bonaparte. Leer felt confident enough to assert that at long last the principles of military art, regardless of the format in which they were practiced, had proven themselves to be eternal and unchanging. Admittedly, Moltke's conduct of warfare differed greatly from Napoleon's in that he had taken advantage of new techniques such as the railroad and the telegraph to move his troops to the most advantageous point at the most advantageous time but this merely proved the efficacy of operational lines. New "physiognomies" might emerge in the conduct of warfare, but the essence of the laws did not change.[113]

By 1873 Leer's success with the criticohistorical method allowed him to restructure the lecture on the history of military art, this time abolishing the now proven arbitrary distinction between offense and defense which he had hitherto opposed but was obliged to include in his course. The faculty of military art was now to be conducted by means of the criticohistorical method; it would review historical data in conjunction with subjective speculation of recent military operational experiences of both antagonists in the recent Prusso-Austrian War of 1866 and the Franco-Prussian War of 1870/71.[114]

However, during the years directly after the Franco-Prussian War, works of various other instructors in the faculty of military art were to take a significant part in the further development of military theory in Russia. In 1875, when Berens was forced to retire due to poor health, his course was taken over by two extremely talented and ambitious instructors, Colonel Stankevich and Colonel Gudim-Levkevich, who took on three new adjunct-professors as assistants: Captains Sukhotin, Puzyrevskii, and Shenshin.[115]

Responsible for writing the first text in 1867 on the *History of Military Art* (*Istoriia voennogo iskusstva*) according to the recommendations of the special meeting, Stankevich was by this time conducting a historical analysis of the Russo-Polish War of 1830/31 in an attempt to show the dependence of military art's development on such factors as the composition of the troops, various military establishments, and the interrelationships of tactics with logistics.[116] Likewise Gudim-Levkevich was soon to publish his *Historical Development of Russia's Armed Forces to 1708,* which was the first significant expression of a nationalist trend in Russian military historiography promoting the acceptance of a particularly Russian military art. Claiming that military traditions and institutions unique to the Russian nation made themselves felt in the military art practiced by the Russian army, Gudim-Levkevich provided a new subjective dimension to Stankevich's statistical work: the proposition that military art was determined by the subjective contributions of a nation's entire culture rather than the limited expression of a commander's individual genius. The implications for the study of military history were enormous: both the subject and method by which to study this extended definition of military science's subjective component were to undergo considerable debate in the years to come. The support for recognition of this uniquely Russian military art on the part of these new instructors was evidently significant enough to bring the issue to the attention of the academic conference in 1877. However, before any conclusion could be reached the demands of the Russo-Turkish War (1877/78) transferred many of the instructors of the academy to field duty in the Caucasus, leaving the direction of the curriculum firmly in Leer's hands.[117]

One of the more important and least accessible military-scientific efforts to come out of the new professionalism of Miliutin's military reforms was the rationalization of the strategic planning process. There was nothing new about strategic planning; in one form or another it was a hallmark of the military profession since time immemorial and was the specific dominion of military geography and statistics. During the Napoleonic wars there had been a plethora of plans to cope with the threat of Napoleon's armies, these finding amusing treatment in Tolstoy's accounts of various war councils in his novel *War and Peace*. During the Crimean War Miliutin himself had been involved in strategic planning for the war effort. But even in that instance it was performed on an ad hoc basis. Miliutin had performed his duties unaided and under the title of special consultant to the war minister, and his efforts had been largely ignored by the commanders in the field.[118]

In 1864 Miliutin began to systematize the collection of statistical information on the armed forces of foreign countries to be procured through Russian consulates and ambassadorial missions. Ostensibly for the purpose of "acquainting Russian officers to the armed forces of foreign governments," the statistical review was to be concerned with the following: (1) composition and numbers of troops in peace and war; (2) central and regional command structure; (3) logistical support structure; (4) armaments and weapons of troops; (5) military law and order; (6) military-technical establishments; (7) budget of the war ministry; (8) fortresses and their strength; (9) strategic studies on the main theaters of war; and (10) general conclusions on the strength of the armed forces.[119] This information was then to be sent to the war ministry and the individual military districts which were to use it in formulating their mobilization procedures for their assigned theaters of war.[120]

By the early 1870s this rather traditional statistical approach was beginning to adopt the methods of the Applied School which had been incorporated into the curriculum of the Nicholas Academy in the late 1860s. N. N. Obruchev as president of the newly created military-scientific commission of the general staff had begun to conduct staff rides to study the terrain and disposition of troops on the occidental frontier in 1871. Recognizing the significance of the military-political shift of power brought about by the Franco-Prussian War of 1870/1871 and the implications it would have for long-term diplomatic and military relations in Europe, Obruchev set about developing his concerns for the strategic defense of the Russian Empire against Germany.[121] During the winter of 1872/73 he set these forth in a review of the empire's strategic situation, "Considerations on the Defense of Russia," which sought to provide guidance for the military district staffs in their review of individual theaters of war. In the introduction to this new statement of policy, Obruchev summarized the new international situation as follows:

> All European powers alike are filled with a desire for peace, yet all of them continue to manufacture their armaments: enstrengthening cadres and reserve armies, and preparing for their more rapid mobilization, building new fortifications and strategic railroad lines. In preserving peace, all prepare for war. For this reason we must urgently take measures to preserve the safety of the empire, even more so in an era when the transition from peace to war can be completed instantaneously.[122]

The strategic planning reviews conducted by the military district staffs were to address all the possible combinations of threat which existed within their assigned theater of war, and they were admonished to keep in mind a series of questions to guide them in their work: "1). How strong must our armed forces be to defend the empire? 2). How quickly can they be mobilized? 3). How quickly can they be concentrated against the enemy? 4). Are our frontiers prepared for a successful defence?"[123] Obruchev's own general opinion concerning the answers to these questions was that if the empire were threatened by a coalition of enemy armies as it had been in the Crimean War, or if it were confronted with the armed force of only one of these powers, the result would be disastrous.

The first major aspect of the actual formulation of strategic plans involved administrative measures for the mobilization of the military district's material for military operations, a concept which Obruchev admitted was "a matter which is completely new to us." Citing the Russian army's dismal attempts to mobilize for the Franco-Austrian War in 1859 and the two-and-a-half-month period required to react to the Polish crisis in 1863, Obruchev commended the military district staffs' subsequent improvements in this respect but concluded that they still needed to match the mobilization rates of the neighboring powers. To do this not only the military district staffs but the whole Russian government would have to attend to the development of railroad lines throughout the empire.[124]

Once the forces of a military district had been mobilized, the second aspect of the strategic plan was concentration of these forces in the theater of war, considerations which depended on the distribution of the troops within the empire and the availability of rail transport for their delivery.[125] Given the number of available rail lines, Obruchev asserted that the Russian army could complete its full mobilization and concentration of troops on the western frontier within fifty-four to fifty-eight days as compared to the twenty to twenty-three days Germany would require to fully transport its forces to the frontier. This relative disadvantage in mobilization and concentration rates confronted Russian strategists with a dilemma: either face hostilities with insufficient forces until the army was fully mobilized or give up space in order to gain time to complete mobilization. In either of these two cases the Russian army could count on having lost Poland and most of Lithuania to the Germans, or if fighting Austria would lose most of Poland and the northwest part of Vol'na.[126] Obruchev concluded from this discussion that the basic task of strategic planning for the European theater of war was to prepare for a war against a coalition formed by these two powers.

While recommending that the battleground be contained to the region around the Visla River due to its advantages for the defense, Obruchev suggested that plans should be made to capture East Prussia and Galicia to prevent the seizure of Poland, and that cavalry should be mobilized in the western frontier region for the purpose of threatening the enemy's concentration of forces. Defending Warsaw still remained a problem, and as Obruchev felt fortification would be too costly and would not guarantee the safety of the entire metropolitan population from German or Austrian attack, he advocated withdrawing the battlefield to the region between the Bug, Narev, and Visla rivers.[127]

When war eventually did break out, as it did in the spring of 1877, it was not to the west against the Austro-Hungarian and Prussian empires but to the south against the Ottoman Empire over the self-autonomy of the Balkan nationalities. The Russian army did not find itself wholly unprepared for this contingency. As early as 1866 Obruchev had formulated a war plan for military operations against Turkish forces in the Balkan theater. In promoting the cause of Balkan independence he had to take into consideration the possibility that any renewal of hostilities in the Balkans would eventually entail war with a coalition of European armies, as had been the case in the Crimean War, and that Russia no longer could rely on a Black Sea fleet to support its land operations.

Given these limitations and assuming that any military efforts against Turkish rule would gain local support, Obruchev explored the possibility of concentrating a large contingent of Russian cavalry in Romania which, when having forced the Danube River, would be reinforced by Serbian troops to force its way to the very walls of Constantinopole within a few weeks.[128] When diplomatic crisis arose over the resurgence of the eastern problem in the spring and summer of 1876, the war ministry assigned Obruchev to reevaluate this old contingency plan to encompass simultaneous military operations against the Turks in the Balkans and the Caucasus theaters of war.

Combining his recommendations with more recent analysis performed by N. D. Artamonov at the main staff who had focused on the problem of fording the Danube River, Obruchev drew up a new war plan entitled "Strategic Considerations of the War Plan" which was submitted to Miliutin and the tsar in October 1876. This plan differed from the original plan only in its devotion to more specific detail. The main offensive conducted by strong cavalry in the Balkans theater was to avoid the strong Turkish-fortified quadrilateral at Rushchuk, Silistria, Varna, and Shumla in the eastern Balkans and attack along the more central direction of Sistovo, Tirnovo, and Adrianopol for a timely march against Constantinopole. Despite the lack of naval support the Russian concentration of forces in Romania and occupation of the northern region of Bulgaria would be facilitated by the use of a recently built railroad network. Heartened by the apparent demoralization of the Turkish army due to its role in the brutal suppression of the Bulgarian uprising, Alexander II approved of the new war plan and on November 1 the decree for the mobilization of the Kiev, Odessa, and Kharkhov military districts was formally enacted.[129]

From its beginning the mobilization of troops from these three military districts was encumbered by confusion over the use of the railroads. According to military observers of the Russian mobilization efforts during the winter months of 1876/77, the transition of railroad use from civilian to military transport was marked by poor coordination between the railway companies and a host of government ministries which, in the rush to gain control over various rail lines, resulted in greatly reduced flow of rolling stock. Contrary to the officially esti-

Plan for the Russo-Turkish War of 1877-1878 in the Balkan Theater
As developed by N. N. Obruchev.

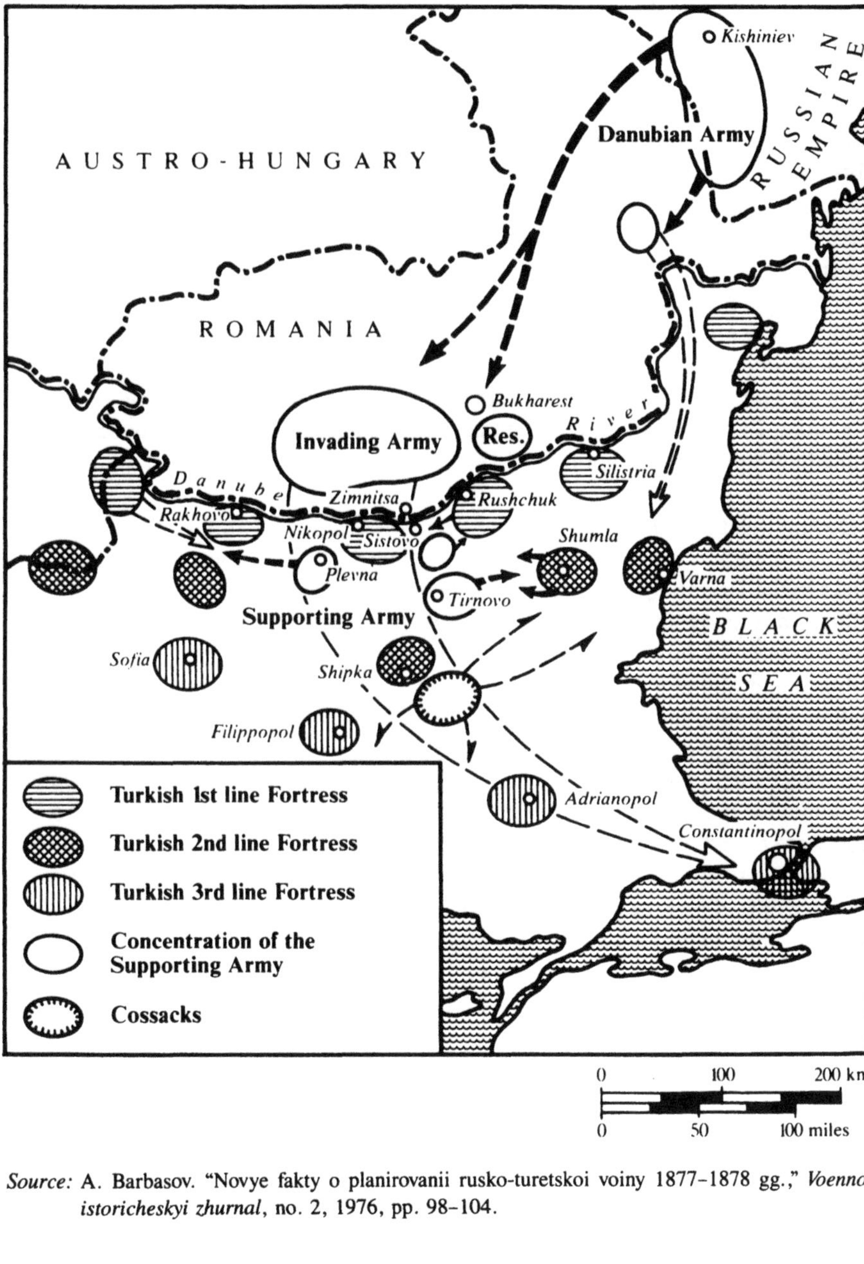

Source: A. Barbasov. "Novye fakty o planirovanii rusko-turetskoi voiny 1877–1878 gg.," *Voenno-istoricheskyi zhurnal*, no. 2, 1976, pp. 98–104.

mated mobilization of four hundred thousand troops to the assembly point in Bessarabia (modern-day Moldavia) by January 1877, independent observers estimated a significantly lower number of 118,000.[130]

During this time Russian diplomacy continued to seek a variety of options toward the resolution of the Balkan problem. During December and January 1877 at the Conference of Constantinopole, Russian diplomats sought to negotiate a wide degree of autonomy for the people of Bulgaria, Bosnia, and Hertsegovina. While these negotiations ground to a stalemate in January, they concluded a secret agreement with Austro-Hungary which guaranteed Austrian neutrality in the event of war with Turkey.[131] Given the new political situation, Obruchev's "Our Political Position at the Present Time" presented to the tsar in February, pressed for immediate military action against the Ottoman Empire to present a fait accompli to the concert of European powers intent on delaying the creation of a strong Bulgarian state.[132] Ignoring these recommendations the tsar continued to seek peaceful negotiations with the Turks, resulting in the highly unstable agreements made in the London Protocol at the end of March 1877. When it became apparent that Britain had become involved in rearming the Turkish army and that Turkish forces were regrouping in northern Bulgaria as well as rebuilding fortresses along the Danube, Alexander II acquiesced to demands for immediate military operations and reluctantly declared war on Turkey on April 17, 1877.

During this time Obruchev had not been idle. In these months of inconclusive diplomatic negotiations the superiority which the Russian army held over the Turks in the Balkan theater had gradually dwindled, and in the face of this loss of military advantage the push for Constantinopole was even more crucial for military success.[133] Obruchev adhered to the fundamental proposal for two theaters of military operations with the main effort to be directed across the Balkans toward Sistovo, Tirnovo, and Adrianopol before marching on Constantinopole. In order to achieve this goal the Russian contingent was molded into two forces: an invading army which was to cross the Danube and force its way through the Balkan mountains at the Shipka pass to threaten the Ottoman capital at the end of three or four weeks, and another army securing its lines of communication in northern Bulgaria.[134] In this way the Russian army would avoid the expenditure of vital resources for eliminating the threat from the fortified quadrilateral in the eastern Balkans while taking advantage of popular support, two options which the Russians had ignored to their detriment in their campaign against the Turks in 1828–1830.[135] All in all, Obruchev's plan called for the creation of a Danubian army which would consist of 186 infantry battalions, 213 cavalry squadrons and companies, and 774 pieces of field artillery totaling 303,000 troops and 90,000 horses against an estimated enemy force of 160,000 troops estimated to grow by another 50 to 60,000 within four to five weeks.[136]

With the declaration of a second mobilization in April, preparations were made to assemble the Danubian army according to the prescriptions outlined in the war plans. Although full assembly of the prescribed forces was not completed by the beginning of June, the first major phase of the plan—the crossing of the Danube and establishment of a bridgehead at Sistovo on Turkish-controlled territory—was entrusted to the 14th Infantry Division under the command of Dragomirov. Preceded by successful diversionary tactics and conducted under the cover of night by troops rigorously trained for the assault, Dragomirov succeeded in crossing the Danube in a matter of hours with relatively few losses, a feat of arms which rivaled similar attempts to force the crossing in past campaigns.[137] Although the Danubian army still lacked the full complement assigned to it in the war plan, it had successfully crossed the first major barrier and was only opposed by a garrison of four thousand Turkish troops at Tirnovo in its push toward Constantinopole.

At this juncture the high command, apprehensive because troops had not yet arrived to cover the flanks of the invading army, chose to depart from Obruchev's plan and sent only General Gurko's forward detachment of ten thousand troops to capture the Shipka pass and stir up national rebellion while Grand Duke Alexander Alexandrovich's Rushchuk detachment and General Kridner's western detachment covered his left and right flanks, respectively.[138] Gurko, still following the timetable set for him in the war plan, quickly neutralized the Turkish garrison at Tirnovo, crossed the Balkans at Khoinkioi pass, cleared Shipka pass, and then prepared for the attack on Adrianopol. Meanwhile, the Turkish forces of Osman Pasha to the west and Suleiman Pasha from the south were converging to threaten the security of this advance and Gurko was forced to retreat.

When Kridner's western detachment struggled through the surprisingly stubborn Turkish resistance at Nikopol to be brought up short at Osman Pasha's hastily prepared defenses at Plevna in early July, it was apparent that the momentum of the Russian offensive had passed. As Kridner concentrated forces for a concerted assault on Plevna in the middle of July, the war minister recommended to the tsar that the Obruchev plan be temporarily shelved in order to organize a defense and call up reserves for an extended siege.[139] At this time Miliutin also approached the British military attaché and observer, Colonel Wellesley, to act as extraordinary intermediary between Alexander II and Queen Victoria in ascertaining how long the war would continue without direct British intervention on the side of the Turkish army.[140] Led to understand that intervention would most likely begin in the spring, Miliutin was now able to redraw the schedule of the war plan to account for the temporary defense. Meanwhile, the first concerted assault against Plevna and the battle of Lovcha to cut off Osman's communications with Suleiman's army had proved to be inconclusive and concerted efforts were being made to assemble what was to become the third and final assault against Plevna.

At the end of August the Russian high command in conjunction with the recent arrival of Romanian forces recommenced the assault on Plevna with an extended artillery preparation which, when it failed to destroy the Turkish defenses after four days, was followed by a bloody and fleetingly successful infantry attack.[141] Here in the frightful expenditure of men and material were the first glimpses of modern attritional warfare, a result of technical advances which was to severely challenge the soundness of Dragomirov's carefully balanced theory of preparation by firepower and assault by cold steel as well as the high command's responsiveness to support tactical successes. Only the determination of tactical commanders, epitomized by the leadership of Skobelev, willing to lead from the front-line in order to estimate the exact moment to bring in reinforcements with tremendous loss of life for both troops and officers, were able to threaten the power of the defensive. In the end when the lack of coordinated control, reconnaissance, and combined arms support failed to take Plevna by storm the Russian and Romanian High Command settled down to a three-month siege which finally ended in late November with Osman Pasha's unsuccessful bid to break out of the Russian defenses.[142]

With neutralization of Turkish forces north of the Balkans the Russian high command, assured of weakened Turkish resistance during the winter months and the lack of any threat from British intervention until spring, chose to resume the strategic directions and timetable of the offensive plan. Within three weeks the detachments of Gurko, Kartsov, and Radetskii had pushed through the Balkans at the Arab-Konak, Trionov, and Shipka passes, respectively, to begin the march on Constantinopole. By early January Gurko's detachment had destroyed Suleiman Pasha's army at Filipopol and Skobelev's advanced guard was to fight the last battle of the war at Adrianopol.[143]

Although Russian military victories had fallen short of achieving the objectives sought by the original plan and were eventually compromised by larger European political interests in the Congress of Berlin, the war with Turkey had proven to be a useful challenge for the reformed Russian imperial army. The combination of administrative regionalization in the forms of Miliutin's new military district system and operational intellectualization in the form of Obruchev's war plans withstood the test of superior defensive technology and tactics to become permanent components of Russian military professionalism.

NOTES

1. G. P. Meshcheriakov, *Russkaia voennaia mysl' v XIX v* (Moscow: Izdatel'stvo 'Nauka', 1973), 86.

2. Ibid., 88. This relationship between military theorists at the Nicholas Academy and the intelligentsia is discussed in N. Makeev's *N. G. Chernyshevskii—redaktor "Voennogo sbornika"* (1950).

3. E. W. Brooks, "The Improbable Connection: D. A. Miliutin and N. G. Chernyshevskii, 1848–1862," *Jahrbucher fur Geschichte Osteuropas* 37 (1989): 34–40.

4. P. A. Zaionchkovskii, *Voennye reformy 1860–1870 godov v Rossii* (Moscow: Izdatel'stvo Moskovskogo Universiteta, 1952), 221.

5. Ibid., 47.

6. A. A. Strokov, *Istoriia voennaia iskusstva* (Moscow: Voenizdat, 1965), 605. See also Brooks, "Improbable Connection," 25.

7. Zaionchkovskii, *Voennye reformy 1860–1870 godov v Rossii,* 222.

8. Ibid., 51. See also Brooks, "Improbable Connection," 35. V. M. Anichkov, professor of military administration, was responsible for writing most of this document.

9. Ibid.

10. Ibid., 54.

11. Ibid., 100.

12. F. A. Miller, *Dmitrii Miliutin and the Reform Era in Russia* (Nashville, Tenn.: Vanderbilt University Press, 1968), 41.

13. Zaionchkovskii, *Voennye reformy 1860–1870 godov v Rossii,* 69. The flexibility of the division was evidenced in increased mobility due to the compactness; troop numbers could be increased or decreased as the situation demanded without disturbing the internal structure of the division or that of the military district. Further discussion of the advantages of the division over the corps structure is to be found in Miller, *Dmitrii Miliutin and the Reform Era in Russia,* 41.

14. Miller, *Dmitrii Miliutin and the Reform Era in Russia,* 49. Others argued that the district commander should take on the full responsibilities of a corps commander. Miliutin argued that the district commander should have the prerogatives of a corps commander but that since he was to remain stationary, his relationship with the tactical commander, who would be a division commander, would be one not unlike that of the war ministry's supervisory role over the military district commander. Thus the tactical commander would retain full executive powers over his respective unit.

15. Zaionchkovskii, *Voennye reformy 1860–1870 godov v Rossii,* 71.

16. Ibid., 74, 77.

17. Ibid., 60.

18. Ibid., 85.

19. Ibid., 85–86. The proposed military districts were as follows: Finland; St. Petersburg; Baltic (Riga); Northwest (Vilna); Imperial Poland (Warsaw); Southwest (Kiev); South (Odessa); Moscow; Kharkov; Higher Volga (Kazan); Lower Volga (Saratov); Caucasus (Tiflis); Orenburg; West Siberian (Omsk); East Siberian (Irkutsk).

20. Ibid., 88. See also L. G. Beskrovnyi, *Russkaia armiia i flot v XIX veke Voenno-ekonomicheskii potentsial Rossii* (Moscow: Izdatel'stvo 'Nauka', 1973), 207.

21. Ibid., 95. In 1862 the Warsaw, Vilna, Kiev, and Odessa military districts were established. They were augmented in 1864 by seven domestic military districts: St. Petersburg, Moscow, Finland, Riga, Kharkov, and Kazan. By 1865 the other domestic military districts were being formed.

22. Beskrovnyi, *Russkaia armiia i flot v XIX veke,* 211. See also Zaionchkovskii, *Voennye reformy 1860–1870 godov v Rossii,* 140.

23. Miller, *Dmitrii Miliutin and the Reform Era in Russia,* 61.

24. Zaionchkovskii, *Voennye reformy 1860–1870 godov v Rossii,* 56–57.

25. Meshcheriakov, *Russkaia voennaia mysl' v XIX v,* 172.

26. E. R. Goldstein, "Military Aspects of Russian Industrialization: The Defense Industries, 1890–1917" (Ph.D. diss., Case Western University, 1971), 46.

27. Ibid., 60.

28. Ibid., 66.

29. Zaionchkovskii, *Voennye reformy 1860–1870 godov v Rossii,* 224.

30. Ibid., 56.

31. Ibid., 228.

32. Ibid., 229.

33. N. P. Glinoetskii, *Istoricheskii ocherk Nikolaevskoi akademii general'nogo shtaba* (St. Petersburg, 1882), 205.

34. Ibid., 209. The provisional regulations stated in broad terms that the new purpose of the Nicholas Academy was to: (1) train officers specifically for the general staff; and (2) spread military knowledge throughout the rest of the imperial army.

35. Ibid., 208, 211. Augmented by war ministry order, Apr. 11, 1863.

36. Ibid., 209.

37. Beskrovnyi, *Russkaia armiia i flot v XIX veke,* 188.

38. Glinoetskii, *Istoricheskii ocherk Nikolaevskoi akademii general'nogo shtaba,* 213.

39. Ibid., 214.

40. S. Wilkinson, *The Brain of an Army: A Popular Account of the German General Staff* (London: Constable, 1895), 60.

41. C. E. White, *The Enlightened Soldier: Scharnhorst and the* Militarische Gesellschaft *in Berlin, 1801–1805* (New York: Praeger, 1989), 98.

42. Ibid., 98–99.

43. Egerton MSS. 3168, f. 44.

44. L. Leonidov, "Iz istorii russkoi voenno-istoricheskoi mysli 60-x godov XIX veka," *Voenno-istoricheskyi zhurnal* 10 (1973): 97.

45. Anon., "Russkoe voennoe obozrenie," *Voennyi sbornik* 4 (1865): 141.

46. Zaionchkovskii, *Voennye reformy 1860–1870 godov v Rossii,* 231. Also cited in Glinoetskii, *Istoricheskii ocherk Nikolaevskoi akademii general'nogo shtaba,* 219.

47. Leonidov, "Iz istorii russkoi voenno-istoricheskoi mysli 60-x godov XIX veka," 98. See also "Russkoe voennoe obozrenie," 140.

48. "Russkoe voennoe obozrenie," 140.

49. Ibid., 142.

50. Ibid. Although the report does not identify Leer or the criticohistorical method by name, the description of this new analysis fits very closely with the discussion in Leer's article "Znachenie kriticheskoi voennoi istorii v izuchenii strategii i taktiki" (*Voennyi sbornik* 5 [1863]):

> In accordance with the general "practical" direction of the entire academy's curriculum distinction must be made between them and the underlying proportionality between elements: historical, critical, and dogmatical; on the basis of these proportions one can recognise the following: that too much development in historical fact consumes much time to the detriment of practical exercises; that criticism, if it is not based on detailed analysis of well-documented fact can accustom the student to superficial opinions and unconditional verdicts; finally, excessive dogmatism in military science can lead to one-sidedness and exclusive theories completely divergent from the direct practical goals of specialized military training.

51. Leer, "Znachenie kriticheskoi voennoi istorii v izuchenii strategii i taktiki," 85.

52. "Russkoe voennoe obozrenie," 143. See also Glinoetskii, *Istoricheskii ocherk Nikolaevskoi akademii general'nogo shtaba,* 250–52.

53. "Russkoe voennoe obozrenie," 147.

54. Ibid.

55. Ibid., 147–49. See also Glinoetskii, *Istoricheskii ocherk Nikolaevskoi akademii general'nogo shtaba,* 223–27; Zaionchkovskii, *Voennye reformy 1860–1870 godov v Rossii,* 232.

56. Glinoetskii, *Istoricheskii ocherk Nikolaevskoi akademii general'nogo shtaba,* 221.

57. Ibid., 228.

58. Ibid., 233, 249. The faculty of military art was officially established in 1869 and consisted of three professors and two assistant professors.

59. Ibid., 234.
60. Ibid., 235–36.
61. Egerton MSS. 3168, f. 36.
62. Carton 7N 1484, A. Nessel, "L'Académie d'État-major Nicolas à St. Petersbourg" (Paris, Feb. 26, 1897). The French observer, writing about his visit to the academy in 1897, noted this cult for the grand man of war in Russia and recounted how one of the professors who had visited St. Cyr was scandalized to find that the French school of higher military education did not exhibit a portrait of Napoleon.

In a letter to Alexander II concerning a portrait to be sent to the Nicholas Academy, Jomini expressed his approval of having his portrait hung in the main hall alongside that of Medem whom he accredited as the "first to provide a solid and rational base in the instruction of the military sciences in Russia" (Egerton MSS. 3168, f. 55).

63. This had changed by 1897 (Carton 7N 1484, Nessel, "L'Académie d'État-major Nicolas à St. Petersbourg").
64. Ibid.
65. Ibid.
66. L. Leonidov, "Voennyi istornik A. K. Puzyrevskii," *Voenno-istoricheskyi zhurnal* (1969): 92; Glinoetskii, *Istoricheskii ocherk Nikolaevskoi akademii general'nogo shtaba,* 253.
67. Carton 7N 1465, Colonel Gaillard, "Sur la liberté de discussion en matière technique dans l'armée russe" (Sept. 10, 1874).
68. Ibid., 262.
69. A. Avdeev, "Voenno-istoricheskie issledovaniia v russkoi armii," *Voenno-istoricheskyi zhurnal* 3 (1986): 81–82. See also Glinoetskii, *Istoricheskii ocherk Nikolaevskoi akademii general'nogo shtaba,* 259. Bogdanovich was made an honorary member of the academic conference as well as taking part in the special meeting's reevaluation of the curriculum.
70. Glinoetskii, *Istoricheskii ocherk Nikolaevskoi akademii general'nogo shtaba,* 262.
71. A. Avdeev, "Voenno-istoricheskie issledovaniia v akademii general'nogo shtaba russkoi armii," *Voenno istoricheskyi zhurnal* 12 (1987): 77.
72. Leonidov, "Iz istorii russkoi voenno-istoricheskoi mysli 60-x godov XIX veka," 99.
73. Glinoetskii, *Istoricheskii ocherk Nikolaevskoi akademii general'nogo shtaba,* 260–61.
74. Ibid., 273.
75. Ibid., 274.
76. B. W. Menning. "Bayonets before Bullets: The Imperial Russian Army, 1861–1914" (unpublished manuscript, 1989), 213.
77. G. A. Leer, *Opyt kritico-istoricheskogo issledovaniia zakonov iskusstva vedeniia voiny* (St. Petersburg, 1869), iii.
78. Ibid., i.
79. Ibid., ii: "Only strict and harmonious essays of history and theory, i.e., historical and theoretical (philosophical) means of expression complementing each other: in other words, the critico-historical method of research and presentation of science is to place it on a firm, positive footing."
80. Ibid., 2–3.
81. Ibid.
82. Ibid., 18–19.

83. Ibid., 7.

84. Ibid., 8. Later in the 1890s one of Leer's students, E. I. Martynov, was to amend this orientation to the effect that the tremendous growth in complexity of military affairs by the 1890s had mitigated Napoleon's maxim concerning the human factor, placing more emphasis on the preparation and logistical administration of warfare requiring intellect, and less on the instinctive genius of the commander.

85. Ibid., 14.

86. Ibid., 16.

87. Ibid., 22–23.

88. Ibid., 23.

89. Ibid., 27.

90. Ibid., 28.

91. Ibid., 29–30.

92. Some of the military theorists he analyzed in his works were Jomini, Bulow, Lloyd, von Clausewitz, Willison, Dekker, Napoleon Bonaparte, Verngorst, Medem, and Bogdanovich.

93. Ibid., 38.

94. Ibid., 39. Leer later came under attack on that point (idealism and obscurantism).

95. Ibid., 73. See also Meshcheriakov, *Russkaia voennaia mysl' v XIX v* 191. Leer's concept of the plan of war did not include a mobilization plan, an omission which showed that Leer had not appreciated the importance of preparation in conjunction with the conduct of war.

96. Meshcheriakov, *Russkaia voennaia mysl' v XIX v,* 189.

97. Menning, "Bullets before Bayonets," 42.

98. Ibid., 44.

99. M. I. Dragomirov, *"Guerre et Paix" de Tolstoi au point de vue militaire* (Paris: Boudoin, 1896), 1–2. See also Carton 7N 1484, Captain P. Dreyfus, "Étude sur l'armée russe. Le Général Dragomiroff" (Perpignan, May 26, 1896).

100. Dragomirov, *"Guerre et Paix"* de Tolstoi, 78.

101. Ibid., 4.

102. Ibid., 78.

103. Ibid., 88.

104. Ibid., 60.

105. Menning, "Bayonets before Bullets," 51.

106. Symbols were very important to Dragomirov because they represented tangible icons in important abstract concepts which needed to be constantly reinforced in the minds of the troops: "Symbols have even more power for 'grand individuals,' the collective—the regimental banner as tangible evidence of solidarity" (Dragomirov, *"Guerre et Paix" de Tolstoi,* 74).

107. Menning, "Bayonets before Bullets," 51.

108. Dragomirov, *"Guerre et Paix" de Tolstoi,* 85.

109. Menning, "Bayonets before Bullets," 49.

110. Ibid., 53–54. Throughout 1870/71 Zeddeler gave lectures on the significance of the Franco-Prussian War to the conduct of modern warfare, attracting large audiences including the emperor himself (Carton 7N 1465, General Le Flô, French ambassador to Russia, "Opinions russes sur le procès Bazaine" (Dec. 25, 1873).

111. Some early criticism of his method was voiced by S. Zhilinskii in his article, "O metode issledovaniia i izlozheniia voennogo iskusstvo," *Voennyi sbornik* 6 (1876): 215–53.

112. Avdeev. "Voenno-istoricheskie issledovaniia v akademii general'nogo shtaba russkoi armii," 77.
113. Strokov, *Istoriia voennaia iskusstva*, 630.
114. Glinoetskii, *Istoricheskii ocherk Nikolaevskoi akademii general'nogo shtaba*, 261.
115. Ibid., 263.
116. Leonidov, "Iz istorii russkoi voenno-istoricheskoi mysli 60-x godov XIX veka," 100.
117. Avdeev, "Voenno-istoricheskie issledovaniia v akademii general'nogo shtaba russkoi armii," 78. See also Glinoetskii, *Istoricheskii ocherk Nikolaevskoi akademii general'nogo shtaba*, 258–59.
118. E. W. Brooks, "D. A. Miliutin: Life and Activity to 1856," Ph.D. dissertation. Stanford, 1970, 139.
119. Hoover Institution on War, Revolution, and Peace, voennyi agent (France), 1864, box no. 4, doc. no. 3242.
120. Zaionchkovskii, *Voennye reformy 1860–1870 godov v Rossii*, 281.
121. G. F. Kennan. *The Decline of Bismarck's European Order: Franco-Russian Relations, 1875–1890* (Princeton: Princeton University press, 1979), 44. See also Carton 7N 1465, Colonel Gaillard, "Renseignements divers" (Sept. 14, 1874).
122. Zaionchkovskii, *Voennye reformy 1860–1870 godov v Rossii*, 280.
123. Ibid.
124. Ibid., 281–82.
125. Ibid., 282. In Sept. 1866 Jomini responded to a query about the strategic value of railroads for the defense of the empire by pointing out that in the specific case of Russia's western border there would be a large distinction between commercial railroads and strategic railroads. Building strategic railroads which could be easily destroyed by the enemy was not cost-effective.
126. Ibid., 283.
127. Ibid., 287.
128. A. Barbasov, "Novye fakty o planirovanii russko-turetskoi voiny 1877–1878 gg.," *Voenno-istoricheskyi zhurnal* 2 (1976): 99.
129. Ibid., 101.
130. F. Wellesley, *Recollections of a Soldier-Diplomat* (London: Hutchinson, n.d.), 125–26.
131. Barbasov, "Novye fakty o planirovanii russko-turetskoi voiny 1877–1878 gg.," 101.
132. Ibid.
133. I. I. Rostunov, pod red., *Russo-turetskaia voina 1877–1878 gg.* (Moscow: Izdatel'stvo 'Nauka', 1977), 57.
134. Barbasov, "Novye fakty o planirovanii russko-turetskoi voiny 1877–1878 gg.," 103.
135. D. Reichel, "La position du Général Jomini en tant qu'expert militaire à la cour de Russie," *Actes du Symposium 1982*, Service historique, Travaux d'histoire militaire et de polémologie (Lausanne, 1982), 1:70–71. See also Menning, "Bayonets before Bullets," 74.
136. Barbasov, "Novye fakty o planirovanii russko-turetskoi voiny 1877–1878 gg.," 104. See also Rostunev, *Russo-turetskaia voina 1877–1878 gg.*, 58.
137. See Menning, "Bayonets before Bullets," 78, for a more specific discussion. For a comparison of campaigns, see Reichel, "La position du Général Jomini," 68.

138. *Historical Analysis of the Use of Mobile Forces by Russia and the USSR* (College Station, Tex.: Center for Strategic Technology, Texas A and M University, 1985), 26–27.

139. Barbasov, "Novye fakty o planirovanii russko-turetskoi voiny 1877–1878 gg.," 104.

140. Wellesley, *Recollections of a Soldier-Diplomat*, 138–48.

141. *Historical Analysis of the Use of Mobile Forces by Russia and the USSR*, 29–30.

142. Menning, "Bayonets before Bullets," 103–5. See also *Historical Analysis of the Use of Mobile Forces by Russia and the USSR*, 30–31.

143. Ibid., 32–33.

4

Disputation: Military Science in the Age of Scientific Positivism and Russian Nationalism, 1878–1904

Although D. A. Miliutin successfully enacted his portfolio of reforms throughout the entire spectrum of Russian military institutions during his twenty years as war minister, his actions were continuously challenged by more conservative groups within the military and state bureaucracy. Miliutin's most significant opponent was his former mentor, Prince Bariantinskii, who had intended to reform the central administration of the Russian army upon the model of the Prussian system. Under this system he, as chief of the main staff, would have gained central control over all active troops of the imperial Russian army while entrusting Miliutin with only the general administration of affairs. However, reforms which eventually subordinated the main staff under the war ministry and the creation of territorial military administrations in the form of military districts replacing the traditional corps structure were fruits of Miliutin's labors which Bariantinskii and a growing number of conservative generals found particularly bitter.

Serious and coordinated opposition did not occur until February 1873. Miliutin convened a meeting in order to draw in other ministries to discuss the general strategic and political directions in policy of the empire. Using what was intended to be a first attempt at systematic coordination of national doctrine, Miliutin's opponents launched an overt attack against the central and regional military reforms.[1] In the course of the meetings which followed, Bariantinskii and his aide, General Fadeev, pressed for changes in the system of command by abolishing the military district system and reinstalling the corps structure. In favor with Alexander II, Miliutin was able to hold his ground but did concede the restoration of the army corps in that it would allow the interactive training of all types of troops.[2]

The course of reform changed for the worse when Alexander II was brutally assassinated on March 1, 1881, and Alexander III, who had personally taken part in the opposition against Miliutin's reforms throughout the 1870s, ascended to the throne. Within a month Alexander III had formulated a policy of counter-reform in his manifesto on the strengthening of the autocracy. Miliutin was forced to resign in protest and was replaced by General Vannovskii.[3]

According to his contemporaries and his own self-deprecating remarks, the new war minister possessed personal characteristics which were opposite to those possessed by his predecessor: a man of little education and intellectual talent and of limited energy who suited the conservative motivations of Alexander III. Oddly enough, Vannovskii was aware of his own limitations and the demands of the position. He valued the talents of one of Miliutin's most senior and trusted colleagues, Obruchev, admitting in a request to Alexander III for Obruchev's appointment as chief of the main staff that

> he was not afraid of Obruchev's liberal views and that with this point of view he [did not doubt] that he himself, having spent his whole life in the services, knows very well about administrative affairs and in this sense would not require any help on the part of his aide [Obruchev] but from the point of view of theoretical military considerations and similarly from the point of view of military science he felt that the sole person he could depend upon and who could help advise him on his insufficiencies was Obruchev.[4]

This appointment was not well received by conservatives and subsequent events were to partially justify their misgivings.[5]

Under such seemingly favorable conditions, Miliutin's opposition then took to the field. Alexander III appointed a special commission to examine the reorganization of the military command and control on June 21, 1881, under the chair of General Kotseby. Ten questions were to be discussed by a special commission made up of military personnel from the higher command structure of the army. Along with several members of the royal family were Vannovskii, Obruchev, Dragomirov (as the newly appointed director of the Nicholas Academy, and M. D. Skobelev as an army commander and hero of the Russo-Turkish War, among other participants. By October this commission was prepared to address the ten questions. The first meeting on October 6, 1881, discussed the question: "What problems exist in today's structure of central military administration, specifically the war ministry?" The opposition argued that the existing system confused central military administration with field administration, that it ignored the spiritual side of military affairs, and that it had promoted the growth of bureaucracy to the detriment of efficient administration.[6] The only thing to do was to reform the entire central administration. The other members of the commission, however, were not impressed. Obruchev, voicing the opinion of many of the members, noted that problems did exist but that they did not necessitate scrapping the entire system as proposed by the opposition.

The second question was concerned with how to correctly promote decentralization of command and administration, paring the administrative duties of the war ministry and enhancing the patriotism of the troops. Almost all the members supported keeping the war ministry's existing duties with the possible separation of the chief of the main staff from subordination of the war minister. Eventually, however, the question of whether to keep the war ministry as it was or to split it into two parts was put to a vote, with the majority of the members supporting the notion of the war ministry as a single organization.[7] Alexander III was not pleased by this turn of events and requested the members of the commission to reconsider their resolution—but even this was not sufficient to change the resolution.

After voting in support of a supreme military council which was to resolve future general considerations of military organization and levels of military preparation (but which was not to come to pass), the next meeting of the special commission was concerned with the discussion of the military district system. In discussions over the possible alternatives to the existing regional administration system, Obruchev with the support of Dragomirov set the tone of the commission's eventual decision with the statement that "actually, the problems are enormous but they result from the entire state apparatus and not from the existing system. Emperor Nikolai Pavlovich would not have changed the system if the former had been good. We must, before all, investigate well those faults and then seek the means for improving them."[8] Alternative proposals by the opposition suggesting either the superimposing of larger formations over the military district structure, the rehabilitation of the corps structure, or the adoption of the Prussian system received feeble support. The proposal for limiting the number of military districts while granting more independence to the corps commander and increasing the privileges and significance of the chief of the military district and staff eventually won majority support. This resolution amended but did not fundamentally change many of Miliutin's institutions in spite of the protracted efforts and patronage of the opposition. The counter-reform was unsuccessful because of Obruchev's neutralization of Vannovskii's power to support the opposition, the support of Miliutin's reforms by commanders who had directly experienced the advantages of such reforms, and the fact that fiscal constraints would not have allowed the luxury of yet another reform of the entire Russian military establishment.[9] Another reason was that in contrast to Miliutin's reforms founded on a firm theoretical base, the opposition's attempts at counter-reform lacked a coherent plan.

In the spring of 1878 the director of the Nicholas Academy, General Leont'ev, died suddenly leaving the post open to Dragomirov who was recuperating from a leg wound received during the battle of Shipka Pass the previous autumn.[10] During Dragomirov's subsequent ten-year tenure in this position, the Nicholas Academy experienced a significant increase in reputation in the eyes of the Russian army as a whole, with the works of its professors receiving a far wider audience than they had previously. This period was also marked by significant thought-provoking new works written by an exceptional group of military theorists who, although not necessarily sharing the same opinions about most issues, were allowed to thrive in the intellectual climate Dragomirov was to create for them.[11] This dynamic new era produced collective reviews such as the *Review of Russia's Wars from Peter the Great to Our Times* (*Obzor voin Rossii ot Petra Velikogo do nashikh dnei*) and reference works, such as the *Encyclopedia of Military and Naval Science* (*Entsiklopediia voennykh i morskikh nauk*), both under the general editorship of Dragomirov's colleague, Leer, but which solicited the work of all the professors at the Nicholas Academy as well as some of the more talented students.[12]

However, this did not mean that Dragomirov was not acutely aware of the many deficiencies in the higher military-educational system which had been revealed in the course of the recent war with Turkey.[13] The problem remained very much the same as that which had confronted Miliutin as a newly appointed war minister in 1861: the Nicholas Academy was still training too few graduates deficient in knowledge of the demands of contemporary warfare in an era when general staff officers were gaining importance in the execution of the supreme commander's will. Dragomirov's own experience as a commanding officer in the recent war had shown him the extent of the deficiencies in the command capabilities of the highest-ranking officers, a situation which led him to go so far as to admit that most of the generals were potentially harmful to the effectiveness of the Russian army in both peace and war:

> As before, unfortunately, I cannot help but state that among the senior officers, beginning with the brigade commanders, there are many who are useless in peace time and who will become a threat in time of war. The weak personnel among the senior officers, a significant proportion of whom could not be considered sufficiently trained or capable, in my opinion, demand serious attention.[14]

As a report reviewing the experience of the war was to prove, this was no idle comment: of the corps commanders involved in the recent campaign, five out of twenty-eight had never received any form of higher military education.

In an attempt to address the increasing demand for more students and the inadequacies of the education provided by the Nicholas Academy's curriculum, a special commission was formed in 1879 under Stankevich and Gazenkampf (professors of military administration). The result was reform of the practical exercises in the course on tactics. These exercises were split into two groups, with Leer presiding over the commission to reexamine the regulations for conducting these exercises.[15]

With the sudden death of Stankevich in 1881, the teaching of the history of military art was left to Leer and Stankevich's protégés, Puzyrevskii, Sukhotin, and Maslovskii. The teaching of military administration was performed by Lobko, Gazenkampf, and Colonel Rediger (who later became war minister after the Russo-Japanese War). In military statistics Obruchev, acting as chief of the general staff and honorary member of the academic conference, would occasion-

ГЕНЕРАЛЪ-АДЪЮТАНТЪ

ДРАГОМИРОВЪ.

Adjutant-General Dragomirov

ally lend a hand to Colonel Maksheev (as well as Colonel Kuropatkin, who taught during the academic year of 1879 before being appointed as chief of staff of the Siberian army).[16]

Nor was the military-scientific work to remain solely within the confines of the Nicholas Academy. Study of the experience of the Russo-Turkish War of 1877/78 indicated some problems in the field command caused by the improper transfer of frontier military district staffs to administration of field armies, which compounded the fact that the field duties of troops were becoming more complex with the need to coordinate the construction of field fortifications and to improve communications. To address these immediate practical problems, the main committee created a special commission chaired by Leer and attended by various professors at the academy to study these problems, resulting in the revised field regulations of 1879 and 1881.[17] In 1885 a similar commission—composed of Kuropatkin, Puzyrevskii, and Rediger—was to revise the *Regulations of the Field Administration of Armies in War Time* (*Polozhenie o polevom upravlenii voisk v voennoe vremiia*), working from a code of proposals compiled from a questionnaire which had been circulated among the acting field commanders.[18]

However, these were merely perfunctory changes stimulated by the largest increase of students at the academy since the Crimean War over twenty years before. The tremendous loss of life and inconclusive results of the Russian offensive against Turkish entrenchments at Plevna in late summer of 1877 had indicated significant inadequacies in Dragomirov's theory of tactics and was to restimulate academic debate over the most effective configuration of tactical doctrine to complement new military technologies. In Dragomirov's opinion, however, this combat experience only confirmed the fundamental premise of his original theory—the balance of intellect and will on the battlefield:

> The increasing rarity of bayonet assaults is no proof of the weapon's obsolescence. It only goes to show that one encounters few adversaries dedicated enough to face the bayonet. The use of the "arme blanche" cannot be extended [to dispersed order] because the bayonet and only the bayonet serves as a visible symbol, a manifestation of that supreme resolution, that intense will which attains the goal, the war and everything. Intelligence can but smooth the way.[19]

In maintaining this position, Dragomirov had ignored the specific nature of the unique success Skobelev had obtained at Plevna: a temporary seizure of the Turkish redoubts by relying on close coordination of successive skirmish lines.[20] On the other hand, the experience at Plevna was to inspire the original arguments of his opponents. Led by L. L. Zeddeler this group, later to become known as the Plevna School, reasserted a tactical doctrine which ignored Dragomirov's poetic theory of tactical balance for a more prosaic reliance on brute firepower. Their arguments were persuasive: mindless acceptance of large losses had a very bad effect on morale and could be avoided by initiating an offensive with large amounts of artillery fire in preparation for open formation infantry assault. If the assault was successful, there was to be no pursuit of the enemy – only reassembly of combat order for a successive assault. Such arguments were clearly antithetical to Dragomirov's tactical doctrine.[21]

When the committee on structure and training of troops reconvened in 1879 to discuss possible changes in tactical doctrine for the publication of new field regulations in 1881, Dragomirov's system remained intact. The only concession to the opposition was to elevate rifle fire of skirmishers to equal status with closed formation bayonet assault and to require infantry training to emphasize wider sapper training (digging trenches, etc.) with night attacks (taking advantage of darkness for improved cover). A later theorist at the academy, N. P. Mikhnevich, concluded that "before as now, the bullet and the bayonet will act in unison, although at present the bullet has shown itself to be far from a 'fool,' while the bayonet is a 'fine lad' only under certain [*izvestnye*] circumstances," alluding to V. A. Suvorov's famous aphorism.[22]

For the remainder of his military career Dragomirov was successful in preserving his preconceptions of the military ethic as a basis for tactical instruction. Having secured control over the instruction of tactics at the academy and predominance over the committee on structure and training of troops, he conducted a series of lectures at the frontier military districts in 1878 and 1879 to spread the influence of his program throughout the Russian army.[23] He also took advantage of Miliutin's institutionalization of annual training maneuvers and exercises to develop a program of training composed of new evolutionary training techniques ("extreme" exercises) which would fuse in the mind of the infantry soldier during battle. The purpose of these exercises was "to develop the troops' habitude for maneuver and the '*coup d'oeil*' of the officers so that available resources of terrain and circumstance can be adapted according to regulations."[24]

The first of these exercises, executed on an experimental basis during the summer maneuvers of 1886, was a combined arms detachment pursuing a tactical goal while under live artillery fire. The second exercise was the *Traversant maneuver à la Souvoroff,* which consisted of the systematic movement of infantry ranks through cavalry ranks. The third and most complicated exercise was the small maneuver of two detachments pitted against each other. Each detachment consisted of one infantry battalion, one cavalry squadron, and one battery of field artillery in varying strengths.

After a general officer had been appointed as the referee for the exercise, a short hypothesis indicating the goal, the location of assembly, and specific composition of the detachment was distributed to the two opposing commanders a week in advance to allow them to become familiar with the terrain and to make necessary preparations. A standard questionnaire was also distributed to each commander with the intention of helping him and his subordinate officers in solving problems. For the commander of the offensive side, it posed the following questions:

> 1.) How would I defend the position which I am charged with attacking? 2.) Why in this way and not in some other manner? 3.) What is my order of combat for the attack? 4.) How am I to hide my intentions? 5.) To which of my flanks must I be the most attentive, and what measures will I take to accomplish that? 6.) What orders will I give to my subordinates charged with covering my flanks? 7.) What goals should I assign to different parts of my order of battle? 8.) Where will I place myself [during battle]? 9.) What do I do if the opponent takes the offensive on front, flanks, or rear?[25]

A similar questionnaire was given to the defensive player. The referee would stop the proceedings if serious mistakes were made on either side during the course of the maneuver and would make sure that orders were clear, concise, decisive, and hierarchically transmitted. At its conclusion a critique of the entire proceedings would be made by the referee in the presence of all the officers involved.

The second part of training, the annual grand maneuvers, was to simulate the concert of effort necessary in actual wartime. As in the extreme exercises, the purpose was to solidify the mechanics of maneuver; the maneuvers were conducted in much the same way as the interactive detachment exercises except on a far larger scale, often combining the forces of several military districts. These annual exercises would often reflect regional geographical constraints and personal preoccupations of the arbitrator. When Gurko, commander of the Vilna military district, held the exercises the emphasis would lie on cold weather maneuvers and the use of raiding and reconnaissance cavalry detachments; Dragomirov would emphasize the practice of the bayonet charge.[26] Throughout the 1880s and 1890s these maneuvers proved to be occasions for experimentation with new combat formations and technologies and often contributed information on combat effectiveness to strategic planning.[27]

The tendency to perpetuate preconceptions in the conduct of both the exercises and the grand maneuvers was also to be found in the lecture hall at the Nicholas Academy. During the 1880s and 1890s Leer still dominated the theoretical discussion of tactics, much to the evident impatience of the younger lecturers, Gudim-Levkovich, Puzyrevskii, and Sukhotin. He had integrated his criticohistorical method into the academy's teachings. The core of the course still consisted of the resolution of tactical problems on a map, but was amended by a preliminary enumeration of the applicable principles inherent in the given hypothesis and how these principles would interact with circumstance. Chance still commanded Leer's respect and the many different historical permutations it had caused in the battlefield manifestation of a principle would elicit detailed discussion. Regardless of the extent of detail covered in lectures, however, the agenda for the theoretical discussion was to already be severely limited by adherence to these predetermined principles. New approaches tended to be disregarded for lack of time.[28]

This then was the state of tactical military theory in the Russian army before the Russo-Japanese War of 1904/5. As military theorist A. A. Svechin adroitly characterized this era of Russian military theory, "order was attained at the expense of vitality."[29] Of all the innovations brought into the Russian army by the reform movement, the attempt to formulate a holistic and unifying theory of military art through the balance of subjective and empirical elements of military affairs was the first to fail the test of time. Although in hindsight the lessons of the Russo-Turkish War of 1877/78 clearly suggested the need for a significant reevaluation of strategy, tactical doctrine, military administration, and the impact of new military technologies on other aspects of military affairs, the human element still remained the primary consideration. Another war would have to be fought and lost before Russian military theorists would discern a new direction in the development of modern warfare.

In 1889 Dragomirov took command of the Kiev military district, leaving his post as director of the Nicholas Academy to his long-standing colleague, Leer. In the intervening years Leer, as a prominent professor in the faculty of military art at the academy, had made an enormous contribution to the development of military science – many individual theoretical works elaborating various issues as compilation and editorship of the *Encyclopedia of Military and Naval Science* which was published between 1883 and 1897. Although Leer's influence was great both within the Nicholas Academy and internationally, his work did not escape criticism for its increasingly pedantic style, partiality to Western European military theory, and deductive use of historical examples drawn from Napoleon Bonaparte's campaigns.[30] Nowhere was there better proof of his influence on the curriculum on strategy and his fascination for deducing military theorems from events made obsolete by over half a century than the publication of applied exercises conducted by supplementary course students under his tutelage in 1887: the *Synopsis of Strategy* (*Konspekt strategii*). Drawing extensively from Leer's *Strategiia,* the recognized text on strategic theory, the compilers of the *Synopsis of Strategy* continued to pay homage to the Jominian study of strategy as the "grand tactics" manifested by the genius of the commander within the theater of military operations based almost exclusively on the proper configuration of the traditional lines of operation while ignoring problems of strategic mobilization and material preparation.

The primary concern of these students' exercises was the study of principles governing the main axes (lines of operation) of an army's movement toward obtaining its military goals. These principles emphasized that an operational line must lead to the attainment of the chosen main goal, be convenient, and be safe.[31] The purpose of the exercises, as stated by the final-year students, was to demonstrate

> the full logical and sequential development of each operation with the relatively correct resolution of the following questions: the arrangement of goals, the review of [possible] directions, the strict concordance between the arrangement [of goals] and means for their attainment, the expediency of all separate parts making up the operation, and its internal unity and safety.[32]

In addressing the proper choice of a main military goal, the compilers observed that "the strategic significance of a point has two aspects, 1. the point is important in and of itself (capital cities, fortresses, etc.) or of geographical importance, 2. the point itself is not significant but receives importance in relation to the situation of the army" (a significance characterized by maneuver), thus covering the full typology of possible strategic goals.[33] Using examples taken predominantly from campaigns of the Napoleonic era, the compilers illustrated how in theory these choices were to be made, emphasizing that where a goal was determined by its geographical characteristics, its significance never waned and it could be studied by military geography. When a goal was necessitated by maneuver, however, the inherent complexity of circumstances required study by critico-historical analysis.

Safety of the chosen operational line was the next principle to receive attention, with two aspects of an army's operational line requiring protection—the path taken by the active forces in engaging the enemy, and the movement of supplies and reinforcements to the field forces. Once again relying on Napoleon I's campaigns, the compilers demonstrated that both vulnerabilities of the operational line could be secured by dispatching a strong screening force to protect the side threatened by the opponent, the coordination of column movement, the rapidity and secrecy of movement and control, and avoidance of overextension from supply depots.[34]

Reliance on the legacy of the Napoleonic era was then extended to a strategic review of two noteworthy Napoleonic campaigns, beginning with an appraisal of the overall political situation, the choice of initial goals, a review of Napoleon I's choice of operational lines as well as those of his opponents, and a concluding review of the respective commander's ability to maintain secrecy and quickness of movement along the operational line.[35] In a final flourish of allegiance to the dominant, and by now, traditional progression of an applied strategic exercise, the *Synopsis of Strategy* concluded with a repudiation of various forms of "strategic metaphysics" in much the same style as Leer's earlier works.

The value of this work was that it was an indication of the state of military art as taught in the supplementary course at the time. That it relied exclusively on the Napoleonic military legacy was indicative of Leer's fascination with this era for its historical proximity and complex maneuvers conducted by large, sophisticated European armies while simultaneously examining Napoleon's military genius as a model for emulation by young commanders. That this was felt to be possible and proper was justified by the attitude that "in general theory only gives a starting point for resolving questions in the review of the subject of action: deciding on the actual course of action is the matter of military art."[36] Strategy, after all, gained its full expression in the heat of battle and therefore the study of a positivist strategy was to contribute only indirectly to the development and eventual demonstration of military genius in the commander, not provide a concrete program for action.

Leer's approach to the study of military art was seen by many instructors at the academy and officers in the Russian army to be insufficient. Significant criticism gained momentum directly after the Russo-Turkish War, when Dragomirov's reform of the curriculum at the Nicholas Academy provided an opportunity for the younger instructors who had acquired fresh battle experience in the war to dispute Leer's highly academic and increasingly pedantic courses.

The most expressive of these new contributors to Russian military theory and historiography was A. K. Puzyrevskii who, upon resuming his position as professor of the history of military art at the Nicholas Academy in 1878, proceeded to publish his *Memoirs of an Officer of the General Staff on the War of 1877–78 in European Turkey* (*Vospominaniia ofitsera general'nogo shtaba o voine 1877–1878 godov v Evropeishoi Turtsii*). Intending that these memoirs should have value as a source for future historians, his attempt to provide a semblance of documentary accuracy to the description of events in this campaign provided him with a foundation upon which he was later to establish his independent views on the nature and method of military history.[37]

In March 1881 Puzyrevskii was appointed to take over Stankevich's course on the history of military art for first-year students. During that academic year he relied almost exclusively on Stankevich's lecture notes. The course outline attempted to show the development of military art through an examination of uniquely Russian conditions which directly contributed to the conduct of military operations, that is, the composition and organization of the army, the functions of military administrative institutions, and the means of armed combat without relating them to the genius of the commander-tsar as was traditionally the practice.[38] During this time Puzyrevskii was able to develop his own views for the publication of his second work, *History of Military Art in the Middle Ages* (*Istoriia voennogo iskusstva v srednie veka*), which came out in 1883. In this work he was able to acknowledge his debt to Stankevich's focus by promoting a wider, more materialistic definition of military art. Military art was not to be confined to subjective, archetypal concepts gleaned from the actions of military geniuses alone, but included the socioeconomic conditions of the nation: "Military art consists not only of strategic and tactical operations but is also closely tied to other elements of military affairs and is determined by them; i.e. building up strength, formation, organization, armament, provisions, and education of the troops, etc."[39] This presented the first and most fundamental challenge to Leer's methodology. His attempt at a positive theory to understand the genius of the commander had nothing more to offer than a model for contemplation.

This was not to mean that Puzyrevskii and his colleagues Sukhotin and Maslovskii were to break completely with Leer's philosophical legacy; they only amended it. Puzyrevskii, for one, was to avoid the potential pitfall presented by a wider, more materialist view of history for a purely positivist theory of military art, seen in his acknowledgment that

> in the presence of continual modification of historical circumstances, and consequently the absence of repetition of identical phenomenon in the presence of identical conditions

neither general history nor the history of military art can give a code of laws similar to those of the exact sciences: they can only show the laws of development, each in its own conduct.[40]

Maslovskii, Puzyrevskii's colleague in the search for concrete technique derived from analysis of Russian conditions, also echoed this intent to merely augment the study of eternal principles of military art revealed in the study of military genius with a broadening of the concept of military art to include many other aspects of military affairs: "The contemporary scientific requirements of a completed military historical work, beside the strategic review of a campaign's conduct and description of a battle, must include all the characteristic aspects of military art directly or indirectly influential in the general outcome of military actions."[41] This new objectively oriented military-historical research could only be conducted once a whole new base of information had been assembled, and with this in mind Maslovskii was to devote most of his work between 1889 and his unexpected death in 1894 to the collection and systematization of hitherto inaccessible primary sources related to Russian military affairs over the centuries into new archives and published collections.[42]

Puzyrevskii's work was to address these new criteria for military-scientific research by accepting the already existing distinction between the history of wars and the history of military art. Focusing the brunt of his theory on the former, rather than Leer's preoccupation with the latter, he felt that

such a division is a necessity because the History of Wars claims to show the process of perfecting different aspects of military affairs (the study of an army's composition, organization, fighting ability, armament, and system of supply) at the same time that military art, based on eternal principles, remains unchanged.[43]

Puzyrevskii's attempt to demonstrate the operation of processes within military institutions and other aspects of military affairs allowed him to transcend Leer's historical defeatism. He was thus able to offer an explanation of new issues in military affairs and their significance in the contemporary era as well as their possible future development. This attitude he shared with military historian and fellow theorist of war games, A. P. Skugarevskii.[44]

In this new scheme of historical inquiry the study of the history of military art was to take on the synthesis, or show the interdependence, of these various processes. Aware that such a task would be horrendous in scale, Puzyrevskii, like Medem and other military historians before him, recognized that the focus of such history would have to be confined to a narrow range of issues: "It is sufficient and fully expedite to study only those events, those institutions and ideas which show the general way in the development of military art."[45] Hence the reason for the preoccupation with the development of a uniquely Russian military art, the alternative to the Western European-oriented theory of military art indulged in by Leer and his overly academic protégés.

Gaining inspiration from Golitsyn's theory of the main path in the development of a universal military art in which research would be limited to "that nation, that state which had the fullest expression of military art for the given era,"[46] Puzyrevskii wrote his third book, *Development of Continuous Regular Armies and the Composition of Military Art in the Century of Louis XIV and Peter the Great* (*Razvitie postoiannykh reguliarnykh armii i sostoianie voennogo iskusstva v vek Liudovika XIV i Petra Velikogo*), completed under the sponsorship of the director of the Nicholas Academy, Dragomirov, in December 1888.[47] In this work he sought to demonstrate via a systematic, comparative analysis of Russian and Western European seventeenth-century military art that the centralization of political power in Western and Eastern European countries at that time contributed to the enormous growth in the size of armies which, in turn, inspired the creation of new trends in military art and the birth of military philosophy at the beginning of the eighteenth century.

In many respects this work approximated Leer's own work analyzing the genius of Peter the Great during the Northern War. Puzyrevskii, in fact, openly acknowledged the value of Leer's work. However, the remainder of the work was to traverse an altogether different direction than Leer's teaching. Whereas Leer argued that Peter the Great's genius lay in the ability to borrow Western European models of military administration and adopt Western service regulations and modes of conducting warfare, Puzyrevskii maintained that even though the military power of Russia grew under Peter the Great with the help of admittedly imported Euorpean models, it was also influenced by other factors which he called "national rules."[48] One of the most important factors of these national rules was the individual genius of Peter the Great, but his role was one of insightful synthesizer rather than undiscriminating imitator:

> In his instructions and directions given to the army Peter, for the most part, had a higher concept of military art for that time. Borrowing forms from Europe, Peter, with his genius and independent intellect, could be no imitator. Petty pedantry, one-sidedness and artful methodism so dominant in Western European tactical and strategic concepts were thrown out by Peter who demanded action in accordance with circumstances and not conventional, normal operations.[49]

In this interpretation of the uniqueness of the origins of Russian military art, Puzyrevskii was supported by Maslovskii in the espousal of a distinct Russian military art: "A critical review of the main events up to Peter repudiate the opinion that from the first origins Russian military art was only an imitation of foreign forms: origins in the Norman [invasions], then the peoples of Turkish tribes, Hungarians, Tartars and so on."[50] Maslovskii went so far as to argue that national rules had existed before Peter the Great and offered as further proof of Russian superiority in military art the observation that the troop regulations of 1755 were less inhibited by routine in comparison with Western European regulations of that era.[51] He, like Puzyrevskii, conceded that Peter the Great did reorganize the

Russian army according to models found in the West. Yet this was not a case of blind borrowing but a highly selective use of Western models to suit preexisting Russian needs:

In reorganizing the military affairs of the Moscow government by lifting them up to the level of the military art of Western European armies this great commander introduced new order gradually after a strenuous evaluation of their accordance with historical conditions of Russian life while always conforming to strict implications of battle experience and striving in all cases to accomplish his design with the Russian people.[52]

Although this initial refutation of Leer's teaching was itself highly academic and belonged to a historical era well removed from the contemporary era, it did inspire a new level of inquiry in Russian military thought which was to greatly enrich its further development, namely, renewed interest in collecting, analyzing, and preserving original documents pertaining to Russian military history as well as widening the discussion of issues in military affairs.

When Dragomirov stepped down from his position as director of the Nicholas Academy in 1889 to assume the position of commander of the Kiev military district, Leer took over as the director of the academy to complete his preeminence over the development of the academy's curriculum—or so it seemed. However, the work of Puzyrevskii, Maslovskii, and Sukhotin had progressed to such a degree of sophistication that the pressure they brought to bear on the academic conference for the establishment of an independent course on the development of Russian military art could no longer be withstood. In 1889 Leer as the new director was to add his support for the new discipline in a petition to War Minister Vannovskii, in which he expressed his opinion that

for the last twenty five years Russian military art has been wholly neglected. The other departments of the Academy have achieved a final equilibrium. Russian historical science has made a significant step forward and works of professors at the Academy have established fundamental views on the method of studying military science and on research goals into military phenomenon. Under such circumstances the history of military art in Russia as a subject for instruction must be reworked anew.[53]

This half-hearted concession to his critics was also a signal of a change in the orientation of Leer's own work, namely, to address the expansion in the parameters of military art from the expression of an individual military genius to the cultural development of an entire nation. Leer's reevaluation of his theoretical priorities was not to mean, however, that he was to adopt the arguments used by the younger generation wholesale. On the contrary, he was soon to address new issues in the publication of his essay *Complex Operations* (*Slozhnye operatsii*) in 1892, the conclusions of which formed the content of a lecture he gave at the Nicholas Academy in early December 1893. This essay, published in *Voennyi sbornik* under the title of "Comparison of a Few Evolutions in Strategic Art," was to embroil him in renewed controversy with these critics.

Leer was forthright in his acknowledgment that the technical means of conducting warfare had changed dramatically since the Napoleonic era and a thorough reappraisal of the forms of strategic art was mandatory—forms which he had earlier promoted with pedantic rigidity. The enormous numerical growth of modern European armies coupled with the rapidly growing destructive power of modern weaponry, the geographical expansion of the theater of military operations induced by these huge armies, as well as the development of telecommunications and the subsequent growing complexity of the rear support services indicated that reliance on the historical legacy of the Napoleonic wars would no longer be sufficient. However, this still did not mean that the fundamental, eternal character of warfare had changed.[54] Leer conceded that European mass armies would require the delegation of command responsibilities to lower levels of the hierarchy thus, in effect, removing the supreme commander from direct battle command and making his role more supervisory.

This was a major change from Leer's previously total emphasis on the genius of the commander, but not a major concession. The delegation of authority to tactical commanders would certainly be required for the mobilization and deployment of a mass army, the management of the rear, and the conduct of secondary military operations against the opponent. Ultimately, in the conduct of the main operations these commanders would still submit to the "creativity of the senior commander, those bearers of 'spiritual principles' upon which three-quarters of success depends in war."[55]

These first tentative steps toward an updated study of military art were taken by several of Leer's protégés and students at the academy. One of the more gifted of these young military theorists was P. A. Geisman, a former student of Leer who had been appointed to take over the course on the history of military art left vacant by Puzyrevskii when he was appointed chief of staff for the Warsaw military district in 1889. Within a year Geisman was able to prepare his introductory work, *Short Course in the History of Military Art in the Middle and Modern Centuries* (*Kratkii kurs istorii voennogo iskusstva v srednie i novye veka*), for publication. This course contained a very interesting amalgam of attitudes toward the study and nature of military history. As any military theorist at this time would freely admit, Geisman considered history to be a rich and indestructible "treasure-house" for all aspects of the theory of military art,[56] but the unusual aspect of his approach lay in his opinion that its study could benefit from both Leer's and Puzyrevskii's methods of inquiry.[57]

His view of the fundamental dynamic in military affairs was the same as his teacher, namely, that of the subjective element, the genius of the commander,[58] which inevitably led to the traditional assertion that "the military . . . cannot give unconditional laws; it can only give principles or eternal basic rules which have, do, and always will act as a basic guide in the resolution of questions underlying military art as a practical affair."[59] However, this did not dismiss the opinion that the history of military art was not governed by special codes or principles of historical development common to the study of general history as Puzyrevskii had formerly suggested. The convenience of this orientation to history—and military history in particular—was that it allowed the study of military issues in their ever-widening social context,[60] thereby promoting Puzyrevskii's predilection for a comparative temporal and cultural method. He believed that "in the attainment of the most full and well-rounded conclusions it is necessary to study in parallel the history of military art with the history of the growth of the nation and society," a position which came very close to accepting the existence of national rules so dear to Leer's detractors.[61] However, on the issue of whether Russian military art was unique in its content and development Geisman was clearly of the same opinion as Golitsyn, Leer, and Bobrovskii, that perhaps the single most important influence in the development of Russian military affairs was that of German military art in the seventeenth and eighteenth centuries and that with the reforms of Peter the Great "Russia learned from Western Europe and imitated much from it," a comment which immediately drew sharp criticism from Puzyrevskii.[62]

In 1894 military-scientific research by two of Leer's protégés—Geisman's *Attempt at a Study of the Tactics of Mass Armies* (*Opyt issledovaniia taktiki massovykh armii*) and Martynov's *Strategy in the Era of Napoleon I and in Our Own Times* (*Strategiia v epokhu Napoleona I i v nashe vremiia*)—were to represent a significant contribution to the development of Russian military thought. Geisman's work resulted from his highly original and demanding new course at the academy which sought to address the specific problem of modern, multimillion-strong European armies: the operational manipulation of enormous military power. With this text as an aid for teaching the new course, he addressed in detail the dilemma which confronted these armies, the increasing difficulty of coordinating the application of force according to the dictates of terrain, the opponent's disposition, and circumstance. The sheer enormity of a modern army's size was to reassert the need for balance between the subjective quality of the commander and troops on the one hand and the changing nature of an army's material potential on the other. This was an old equation which recognized the need to neutralize a new trend in military affairs—"materialization": "This crude weapon in its aspect as an unwieldy, awkward army is a problem if it finds itself under the direction of an inadequately skilled commander, but in the hands of a great master of [military] art it can lose, to a certain extent, its negative characteristics."[63]

In readdressing this issue of balance, Geisman adopted the priorities of Puzyrevskii and his colleagues. He argued that future military science should dispassionately evaluate the experience of foreign armies while researching the minutest details of the theoretical, organizational, and operational command of the Russian army in order to develop appropriate techniques for maneuvering modern armies.[64]

The most significant of these techniques concerned the method by which the supreme commander would exercise his will over the entire army. In what was to be the inverse of the concept of *glazomer,* Geisman introduced *orientirovaniia* (orientation) and attributed it to Leer.[65] Since the physical dimensions of modern armies promoted a disintegration in the chain of command, a centrifugal (*tsentrobezhnost*) tendency, it was up to the command to reassert the need for unity, or centripetal (*tsentrostremitel'nost*) tendencies which promoted the ideal method of fighting operations and battles.[66] This was to be promoted in three ways:

> A). by means of directives and instructions,[67] B), in self-orientation, that is to say, the awareness of the local situation . . . regardless if they are bereft of orders and direction from above, C), by means of *close* communication between each army and all other independent groups finding expression in the fact that during each minute each of the armies and groups have an accurate understanding of what is going on in the neighboring armies and groups.[68]

In this way Geisman was to implicitly argue for a new way of structuring the organization of forces and the flow of command to conform to a Prussian method suggested by Moltke the Elder in contrast to the Napoleonic/Suvorovian tradition held dear by the proponents of the Russian national school. It was not until after the Russo-Japanese War in 1904/5 that these concepts reemerged to play a significant role in Russian military theory.

Another of Leer's protégés was Captain E. I. Martynov who, after graduating from the Nicholas Academy in 1892, published a monograph entitled *Strategy in the Epoch of Napoleon and in Our Time* (1894) which was to further develop many of the ideas Leer had stimulated at the academy. Motivated by the revelation that "despite the eternal, unconditional nature of strategic theory, strategic art is not static: changes with new circumstances can also lead to a change in its character, *a new strategic age,*" Martynov was to combine the comparative historical method of Puzyrevskii and Maslovskii with the criticohistorical method to compare the military art of Napoleon I with that demonstrated by Moltke in the Austro-Prussian and Franco-Prussian wars.[69] In order to study this apparent change in strategic styles, he accepted the national school's argument that the study of modern strategic art would have to shift the emphasis of its analysis away from the military genius of the commander to a study of an entire nation's cultural sophistication, thereby consciously transforming the theory of military art into a truly sociological study in breadth of analysis.[70] He sought to enhance the reputation of this inquiry by concluding that it would greatly benefit from the study of civilization's own development, a potential shift in the level of analysis which was to stimulate furious controversy in educated military circles up to the Russo-Japanese War and beyond.[71]

Martynov began his discussion with an observation similar to that made by a colleague at the academy, M. N. Mikhnevich, a few months earlier: "Each cultural period has its own form of military art and strategy: the philosophy of war must be in accordance with the spirit of the times and content of civilization."[72] In elaborating this statement Martynov chose as an example the development of military art in the Middle Ages, a favorite topic of Puzyrevskii's previous works, and argued that the development of centralized political institutions in many European countries at this time promoted a more sophisticated culture. This was to manifest itself in a more effective military art than those governments which did not exercise strong central power.

The majority of the study was devoted to a comparison of the military art at the beginning of the nineteenth century with that at its close. In so doing, Martynov chose to analyze what he considered to be the ultimate symbol of military art at the beginning of the nineteenth century: the military genius of Napoleon, a style of conducting war which was filled with love for poetic risk, absence of formal intellectual training, and complete disregard for the technicalities of warfare—in short, the military artist whom Leer had so flamboyantly illustrated in his *Attempt at a Critico-Historical Analysis of Laws of the Art of Conducting War* in 1869. In stark contrast to this Napoleonic style of strategic art was that of Moltke toward the latter half of the century, a style which Martynov described as "cold-blooded, systematic, and thoroughly trained down to the tiniest *technical* detail in each specialty."[73] Martynov hastened to add that while Napoleon's style of military art rested solely in the individual poetic genius of the commander, the nature of Moltke's style of military art sprang from an entirely different source: from the German people at large. Whereas with Napoleon's death the Napoleonic style of military art and the greatness of the French army died with him, Moltke's death did not have a similar effect on the military art of the Prussian army. For Germany, the death of the commander lowered the level of military art, but there was still much residual culture.[74]

This dramatic shift toward a strategic style dominated by technique was seen by Martynov to be due to three major developments brought on by the application of new technologies to the battlefield: the materialization, numeralization, and democratization of military art. He defined materialization in much the same way as Geisman had done—as the perfection and mass production of rifled, breech-loading small arms and artillery. This led to their rising significance as an actual direct tool in combat whereas they had previously been considered useful only as complementary to the use of cold steel or bayonet charge. The use of the railroad and the telegraph also allowed the timely transport of troops and supplies to the theater of military operations, allowing armies to operate in greater comfort for longer periods of time as well as enlarging the potential operational zone and rear.

Materialization led to what Martynov called the numeralization of military art, or the development of mass armies—armies well over the size of the very largest under the command of Napoleon. Numeralization, in turn, gave rise to the most significant development in military affairs, the democratization of military art. Because of the significant increase in the size and dispersion of armies, the supreme commander was physically unable to maintain direct control over the manipulation of forces. As a result his role had changed to one which was exclusively supervisory; in which he gave general orientation to separate unit commanders in relation to the fundamental concept of the direction of military operations. Actual battle initiative, the central theme of military genius, was now practiced only by the lowest commanders in the hierarchy of command.[75]

These factors led in turn to the growing professional responsibilities of the general staff, which further suggested an increasing democratization of military art.[76] Instead of the poetic genius of an individual, military art had come to depend on the cultural aspirations of a group of men which arose from the civilization to which they belonged. Martynov cited Hegel's historical philosophy to illustrate how this growing trend of democratization had evolved toward greater political, social, and economic freedom in European history since Louis XIV. From this conclusion it was evident that the education of the general staff had become all the more important. In Martynov's words, "rational, conscious initiative is needed, without which it would be impossible to control modern armies. This depends on the culture of the people; on its political and social structures. . . . Military training must now enter the cultural sphere, without it the army becomes a lifeless, machine-like mass."[77] However, Martynov gave no specific recommendations for military organization to assuage the growing size, complexity, and destructive power of a modern army. In the observation that "the army and the nation, the soldier and the civilian are moving closer together," however, he implied the possibility of a militia-type army.[78]

Having elucidated these three trends in the development of military science, Martynov attempted to synthesize them into the development of modern European society. The result of this synthesis was a highly pessimistic prognosis for the future, with humanity reduced to a cog in a machine and society increasingly resembling a giant factory. Within such a level of civilization the very definition of the concept of creativity changed from one which had a poetic, artful connotation to one which had a more technical connotation—what to Martynov was to correspond to a fall to a lower level in the concept of military art. Attempting to mitigate the loss of flexibility and unity of command (*edinonachale*) brought about by the technological evolution of society, he could only suggest a broader education for the general staff with the hope that this would provide "not only a wide world outlook needed, but disciplined independence in agreement with the superior command in the creative aspect of strategy"—a conclusion very similar to that made by Geisman.[79]

Martynov's work was to provoke substantial controversy in the military press, the greatest criticism coming from a librarian working at the military archives, A. N. Petrov, who published an article in *Voennyi sbornik* in May 1894 entitled "Tasks of Contemporary Strategy: As Regards the Work of Captain Martynov, *Strategy in the Era of Napoleon I and in Our Time.*" During the early 1890s Petrov had conducted a historical analysis of the Russo-Turkish War of 1877/78 and had developed strong opinions as to the definition and role of strategy in military affairs. Petrov's first and perhaps most fundamental criticism was Martynov's opinion that the history of military art was directly linked to the history of civilization. This was an interesting thesis, but one which Petrov felt was nevertheless incorrect.[80] If this interrelationship were the case, then the logical conclusion was that the more civilized a nation the higher the development of its military art. Petrov disproved this conclusion by citing the example of Britain, a highly developed country whose armed forces exhibited a level of military art incommensurate with its stature.[81] Nor did Petrov think very highly of Martynov's attempt to promote his thesis with the argument that one could demonstrate the direct relationship between the level of civilization and military art by comparing the military art of decentralized, feudal-type governments to that of more centralized governments. This he saw as an incorrect deduction for the reason that "government structure is only one of the elements of civilization, it is not civilization itself."[82] However, the fundamental source of Petrov's criticism lay in the opinion that military art did not originate directly from civilization but from the force of personality (genius) of a person having the means and power to develop forms of military art.

This twist of argument bordering on the hypocritical stemmed not from a disagreement over the composition of strategy but from a difference of opinion over the nature of warfare and how strategy would best reflect that nature. Petrov agreed that an attempt should be made to illuminate known principles and eternal ideals for the conduct of warfare, but also warned that Napoleon was correct to point out that "in war circumstance rules everything."[83] Seizing upon Leer's recent observation that overemphasis on the determination of ideals could easily degenerate into "inventions, preconceived ideas, Baconian idols," he argued that in this case attention in formulating theory should be devoted to the creation of accepted and well-understood givens of circumstance. By studying the nature of circumstance one could better establish the parameters of theory in military affairs without indulging in idolatry. The nature of circumstance was best described as composed of three components—physical (human power), material, and spiritual—which were further defined as geographical (level of culture to be relegated to the concern of military geography), historical identity, the strength of the armed forces both in numerical and morale terms, the government structure of the nation, material means with which to sustain the war effort, the composition of contemporary technology, the foreign policy of the nation, diplomatic art in forming and maintaining coalitions, and the strictly military administrative functions.[84]

In this way Petrov was to establish the main difference between himself and the academists led by Leer by relegating the fundamental role played by military genius and it cognizance of ideals down to merely another component of circumstance. With the publication of this critique he was to become the main opponent of the academic idealists and gained support from other opponents of Leer's school.[85]

Along these same lines Petrov's earlier work on the significance of the Russo-Turkish War of 1877/78 explored the dilemma of bridging the gap between theory and practice, central to the theory of military art. Recognizing that the theory of military art was a child born of peacetime military establishments which could prove obsolete upon the commencement of hostilities, he emphasized the need for practical exercises accurately simulating the nature of warfare: "one must not fully trust reason because 'to err is human.' Therefore one needs an additional regulator to indicate correct conclusions. This must be experience. Only the deductive method of theory and practice can give one a positive value to a given idea."[86] Over the next four years Petrov was to augment his argument over the nature of theory in a series of essays which were to gain an ever-increasing following among officers in the imperial army. In *Questions of Strategy* (*K voprosam strategii*) (1898), Petrov turned his attention to a perennial issue in military theory: whether strategy was a science or an art. He was inspired by the work of a French libertarian socialist, Pierre-Joseph Proudhon, who differentiated that which was merely imitative, demanding only an acquaintance with technique (science), from that which was creative and based on an underlying idea or concept. With a few references as to how this was portrayed in the practice of other arts, such as music and literature, Petrov then focused his attention on the interaction of the two:

> From this it is apparent that there are many connections between art and science. The goal of science is to collect information and derive from it laws; the goal of art (creative) is to regard these laws as material from which to create something new . . . [and] are never represented in a concrete sense, but are to guide the creativity of individuals. For this reason science is the material of creativity and a regulator of its impulses. Creativity without science is a fantasy.[87]

In his early work analyzing the significance of the Russo-Turkish War, Petrov had implied that the dynamics of this interaction were to be deduced from experience.[88]

This interrelation of science with art was not radically different from that which Leer had hypothesized twenty-five years earlier, and Petrov went so far as to assert that Kantian theory of knowledge had addressed this relationship more objectively in comparison to such epistemological systems as that of Fichte's description of the dialectical process in history. On the subject of planning for future wars, made popular through the writings of German military theoretician von der Goltz,[89] Petrov pointed out that Kant had more to say about the nature of scientific analysis of social phenomena; that the criticohistorical method alone was not responsible for theoretical deduction drawing on the empirical fact of the past but was to be further bolstered by the test of practical experience, a sentiment which he shared with Puzyrevskii. The proper sequence in the development of science was defined as: "1.) the study of known facts, 2.) critical analysis of their relationships, 3.) theoretical conclusions deduced from logical analysis of previous experience, and 4.) verification of the theory through practice."[90] Petrov pointed out that there were to be eternal and perpetual principles upon which to base a scientific theory, but, in what was to be the fundamental difference between his scientific outlook and that of Leer, these principles (having a regulatory rather than dynamic nature) had the advantage of not relying on intellect and were therefore independent of human will. These were what he called the natural laws,[91] which he felt could be described as the concepts of time, space, and force.[92] This was not to deny the importance of the commander in the practice of science in his professional obligations, and Petrov was quick to assert that the genius of the commander was a very powerful element of military affairs. After all, "such is the strong link between science, theory, and art. The first develops laws, the second studies their different features and the last applies them to the given situation."[93]

An offshoot of this discussion which Petrov had used in criticizing Martynov's work was that of the interrelationship of quantity versus quality in the composition of the concept of force. For Petrov the concept of force was one which was perhaps a direct manifestation of the interrelation between science and art, and therefore was of central concern. The question of military force and the definition of its components were certainly not new, and in many respects Petrov's attempt to explain it bore a strong resemblance to Tolstoy's discussion of the components of military power in *War and Peace*. In his critical essay, "Tasks of Contemporary Strategy," this interaction between the quantity or quality of troops with the talent of the commander to determine military power was logically explored through the means of a mathematical proof which sought to disprove Martynov's contention that the role of military art was falling in the age of technology.

This affinity with using mathematical notation for describing the interrelationship between components in military theory had been used to great effect by Tolstoy. More recently an attempt had been made by I. E. Maslov, writing in *Voennyi sbornik* in 1888, to combine probability theory with carefully researched deductions from military history: "If with the aid of these two powerful sources of investigation we are convinced in the correctness of a conclusion made from the history of military art, then such a conclusion acquires the value almost of absolute truth and we attain the possibility of knowing the very rationale of a phenomenon."[94] Although Maslov's efforts represented a legitimate attempt to synthesize the powers of mathematical analysis with specific deductions from military history, Geisman's and Petrov's attempts to apply this to far more complex, value-laden concepts produced truisms bordering on the vulgar. When this tendency resurfaced in Russian military-theoretical literature after the Russo-Japanese War, Martynov properly criticized it as a syndrome of conceptual illiteracy in an officer corps riddled by inadequate professionalism.[95]

Petrov also reacted to Martynov's assertion that the rise in technology led to the ever-larger armies, and he took advantage of this social parallel between the factory and its workers to illustrate Martynov's mistaken concept of the effect of technological growth on the concepts of quantity and quality. Arguing the exact opposite conclusion of Martynov, that the introduction of machines in factories actually cut down the number of workers needed and enhanced the potential for a higher quality of production, he proposed that such implications demanded a serious reevaluation of contemporary strategy.[96] Throughout history there had always been a tendency for great commanders to prefer higher-quality, smaller armies over unwieldy large armies. Even in the age of rapid technological development, this relationship between quality and quantity had been preserved to a certain extent and could be shown in the superiority of the American Confederate army under the command of General Lee throughout most of the American Civil War. The improved capabilities of technology also supported such a view,[97] implying that rather than promoting larger armies and a fall in the level of military art as practiced by the commander, the effect of such technological advances as the field telegraph, observation balloons, photography, and railroads in increasing the parameters and complexity of the battlefield would require greater initiative on the part of subordinate commanders. "Consequently, the role of the commander in our time has not only not fallen, but has become more complex due to contemporary technology complicating the level of circumstance, the use of which is defined by the genius of the commander."[98]

Other than his denial of the need for mass armies which Martynov had stoically accepted, Petrov ultimately came to the same conclusions as Martynov: that in the future the scientific training of general staff officers would become more important. In his later work he elaborated on what this "scientific" training would be, which naturally consisted of his thesis for the combination of criticohistorical analysis with the test of experience.[99]

In a final argument against Martynov's conclusions, Petrov was to provide a chillingly prophetic description of the total warfare of mass armies, as well as a description of its alternative, which was to become the main topic of European military theory after the First World War:

> Mass armies spread out over vast distances would require frontal battle—a complete destruction of military art because battle would not be distinct from a battle between rams. . . . The talented commander will always run away from the necessity of that type of battle, [he] could disguise the transport of a significant part of his army to places where it is not awaited by the opponent, [he] could with success operate against his flanks and especially the rear (wide development of actions on the part of partisan units) in an attempt to envelop and destroy [the enemy's] railroads, etc.[100]

The course of this debate between members of the Russian military-theoretical community was to experience added impetus and new direction in the late 1890s due to the publication of an unusual military-scientific work by a complete outsider to the international military-scientific community, a multivolume work entitled *Future War in Its Technical, Economic and Political Aspects* (*Budushchaia voina v teknicheskom, economicheskom i politicheskom otnosheniiakh*), followed in 1898 by a shorter monograph entitled *General Conclusions from "Future War in Its Technological, Economic and Political Aspects"* written by a financier and economist from Warsaw, I. S. Bliokh. Since the early 1870s, when N. N. Obruchev as principal strategic planner for the Russian army had admitted that the defense and fortification of Warsaw would be impossible, the municipal government of Warsaw had authorized a technical committee to devise a plan for the city's civilian defense. Called upon as the president of the Chamber of Commerce to provide suggestions, Bliokh probably had recourse to bibliographic resources at the staff of the Warsaw military district, at that time one of the largest military-scientific centers outside of the Nicholas Academy, which carried him well beyond the issues of civilian defense.[101]

Drawing from a wide range of contemporary European military literature, especially from the works of Leer, Dragomirov, and Puzyrevskii, Bliokh was to take the premise that the character of future warfare—demonstrated by recent trends in the capacity for destruction and complexity of conduct requiring the total exertion of national resources—precluded the use of warfare as a viable political tool. In the introduction to his *General Conclusions,* Bliokh focused his attention on familiar issues and suggested new implications. Commenting on the problems of command and control of mass armies, he supported the concept of democratization of the supreme commander's military art by insisting that unit commanders would most certainly require more independence of action but in consequence would experience higher casualty rates during combat, implying that with the continued duration of the war these casualties would be replaced by less well-trained or experienced officers, thus promoting a fall in the level of military art.[102]

Bliokh was to support this thesis by examining the principles behind peacetime training of officers, arguing that training based on war experience twenty-five years removed would not satisfy the demands of the new battlefield. Furthermore, this inadequacy was inherent in military science, in that strategic exercises and maneuvers essential to the training of officers in peacetime lacked the all-important element of the psychological moment which motivated the conduct of real warfare—a point which he attributed to Dragomirov.[103]

Universal male conscription also posed the horrendous task of arming and training the inducted men. This was a financial strain which even the richest industrial state could not sustain for long. Compounding this problem were the implications of having to draft working men and the reciprocal strain this would place on the productive forces of a nation's economy.[104]

Having illustrated these eclectic conclusions and their implications, Bliokh described the new method he had developed. Noting that there had been no previous attention devoted to the economic aspects of warfare nor any attempt to apply economic (statistical) analysis to these aspects, Bliokh attempted a broad systems-analysis approach to studying the character of future war with a highly original conclusion:

> Further examining the plans for military operations in the struggle between members of the Great Powers on the European continent we use our main conclusions of their relative strengths to quantify the comparative degrees of stability of each government against the destructive influences of war. From this we come to the conclusion that the government for which war is the least dangerous and the least vulnerable is Russia due to its enormous expanse, characteristics of its earth and climate and the most important, the social way of

life of its people living off the land. Russia could conduct a defensive war for several years during which time the Western and Southern governments, relying on their higher level of culture and larger development of industry and trade but with not enough bread to feed its population, could not conduct war for a whole year.[105]

Immediately upon publication of this enormous five-volume work, Bliokh received sharp criticism from Dragomirov in the form of a reprint of an open letter which Dragomirov had written to the author upon having been asked to review the draft manuscript of the work in 1896.[106] Equally strident criticism was to come from Puzyrevskii, who proclaimed that any genuine resolution of such a complex issue was not possible—especially when it relied on a statistical method: "Bliokh, not a military man, has chosen to analyze military affairs as if it was a straight-forward, almost mathematical equation, which every military man appreciates is not the truth."[107] Such criticism from these two military theorists was to be expected, given their past criticisms of similar arguments presented by Leer and his students.

Although no supportive critiques of Bliokh's work were forthcoming from either Leer or Geisman, his work was to prove stimulating to the young lecturers of military administration at the Nicholas Academy, evident from the publication of an article in *Voennyi sbornik* entitled "War and National Economy," written by A. Gulevich. As a prelude to the publication of a more substantial monograph on the same topic, Gulevich wished to promote the value of conducting a similar comparative statistical analysis of the survivability of Russia's national economy in a future war. He was to directly acknowledge Bliokh's work, though noting that one must view his conclusions cautiously.[108] The rest of the article ingeniously incorporated many of Bliokh's observations into the existing theoretical legacy of Leer, Geisman, and Martynov.

Gulevich's point of departure was the thesis that the economic structure of a nation determined its ability to conduct long and stubborn wars—the wars of the future. Like many military theorists seeking to justify a new eternal principle of warfare, Gulevich turned to a brief reappraisal of history and the great works of military literature, in one case arguing that the true source of Hannibal's and Napoleon Bonaparte's success in war actually lay in the superior vitality of the states they represented. He used quotes to similar effect from Jomini's works as well as those of Montesquieu and Proudhon.[109]

In an attempt to combine this socioeconomic element into the existing philosophical assumption that military affairs was a combination of the spiritual and intellectual strength of the commander, the troops, and now the entire nation, Gulevich toyed with the dialectic of historical materialism: "Victory will go to the side which has superior spiritual strength. But the spiritual, equally as the intellectual strength of each nation, is the result of all its historical life, influenced by elements composed of material circumstances interacting with the nation's [economic activity]"—a view of history which was to allow ample room for the national characteristics of military art.[110] In this manner Gulevich was to claim a wider analysis than that of Bliokh, an analysis which was to take into consideration three fundamental factors in military affairs: the organization of the armed forces, the economic life of the nations involved, and international relations in the context of modern war.

Discussing the issue of universal male conscription in conjunction with the development of railroads as the first fundamental factor in this analysis, Gulevich was to emphasize a point made by Bliokh: "Now, when armies are larger due to enlistment of more mature, older men with families and responsibilities this brings the army even closer to the concerns of the economy."[111] However, he, like Bliokh, failed to draw another important implication from the growing militarism of society which was to reveal itself in the Russo-Japanese War five years later: that the national economy's reaction, in turn, would have a direct influence on the morale of the army.

Another prophetic conclusion drawn from this analysis of mass armies was that not only would they be difficult to maneuver, but that in consequence "no war will be a short war and no defeat will be decisive"—a point of view which Gulevich shared with Bliokh and Rediger. The latter had argued the same point in his *Build-up and Formation of the Armed Forces* (*Komplektovanie i ustroistvo voorushennoi sily*).[112]

A review of European national economies was to substantiate Bliokh's claim that sophisticated industrial economies were particularly susceptible to disruption. The formation of political and military alliances, however, would provide a sense of stability to the international order—the age of modern coalitional warfare. In this discussion Gulevich borrowed arguments from Leer's *Strategiia*, asserting that one of the characteristics of coalitional warfare was confusion between allies with a resultant decrease in the efficient use of resources. This provided a weak point for enemy aggression.[113]

In December 1897 both the war minister, General Vannovskii, and his chief of staff, General Obruchev, were to offer their resignations—ostensibly for health reasons. Vannovskii had wanted to retire for quite some time but felt compelled to officiate his assignment, and although the tsar had offered Obruchev the post, at the age of sixty-eight he did not want to shoulder the increased responsibilities of the war ministry. Anxious to take this opportunity to renew the higher military-administrative personnel and place the war ministry in younger, more energetic hands, Nicholas II lost no time in coming to a decision and appointed the rising star of the Russian army, General Kuropatkin, to the post.[114]

A veteran of the Russo-Turkish War, where he had gained a reputation for bravery as Skobelev's chief of staff, Kuropatkin was a graduate of the Nicholas Academy and had taught military geography there for a short time before being posted to Central Asia. Very popular in both military and civilian circles, he was described by a colleague at the time as "a very intelligent, well spoken, and very ambitious person who wants to place his name next to Suvorov or Skobelev and will do anything to get there."[115] Kuropatkin chose for his chief of staff General

Sakharov, former quartermaster general of the Warsaw military district, over General Puzyrevskii who reacted very badly to being passed over for the post.[116]

Upon his arrival at the war ministry, Kuropatkin conducted a policy review of the war ministry's affairs. To help him in the compilation of this review he enlisted the skills of several professors at the academy: Myshlaevskii on the historical goals of the Russian Empire, Zolotarev on the military statistical analysis of the Russian army's military strength, and Gulevich on military administration. Completed and submitted to the tsar in the spring of 1900, this report reasserted the intellectual traditions and policies which Obruchev had preserved in the functions of the main staff:

With the limitations of human understanding it is not possible to look ahead a hundred years, and we cannot, therefore, lay down what our army will have to undertake in the twentieth century; but by analyzing the past and reviewing our present position among the Great Powers of the world it is both possible and essential to estimate the nature of the work that will come before our army in the next few years at least.[117]

From this stance Kuropatkin argued that since the Russian Empire had reached the extent of its aspirations for expansion, preparations for potential future conflicts would be maintained according to Obruchev's model for fighting defensive wars in the west and Far East while retaining offensive priorities in Central Asia. All in all, this report represented a synthesis of all the intellectual traditions practiced at the academy and was later used in Kuropatkin's apologist arguments for the defeat of the Russian army by the Japanese in 1905.

The year 1898 was a year of transition at the Nicholas Academy. Leer stepped down from his position as director, ending an academic career unparalleled in its qualitative and quantitative contribution to Russian military literature. In his place was appointed a professor from the faculty of military art, General N. P. Mikhnevich, who in comparison had published very little at the time of his appointment. This he soon rectified, with the publication of two large monographs in 1898, *The Basis of Russian Military Art* (*Osnovy russkogo voennogo iskusstva*) and *Strategiia,* both of which were to become standard texts at the academy, the second directly replacing Leer's monumental text. The third monograph, *Influence of New Technological Inventions on the Tactics of Troops* (Vliianie noveishikh tekhnicheskikh izobretenii na taktiku voisk), also published in 1898, was a revision of an article he had originally published in *Voennyi sbornik* in 1892. Synthesizing the strong arguments presented by Leer and his opponents into a clear, unified approach, these three works immediately established Mikhnevich as the leading authority on the history of military art.

In *The Basis of Russian Military Art,* Mikhenvich established a theory of military art composed of three elements: the human as an element in the complement of troops and their command; the weapons or material means with which to defeat the enemy; and conscious ideas (the first recognition of a potentially objective intellectual process greater than any one theorist which would contribute to the practical management of all military affairs—including military art).[118]

In approaching the subject of war and military art, Mikhnevich was to accept the opinion of such theorists as Petrov, Puzyrevskii, and Maslovskii, who insisted that although consideration of material means was important, the moral composition of both troops and their commanders was the most important. This military composition was in turn determined by the broad social and historical legacy of their nation. Military art, then, was to be heavily laden with an ideological character only to be revealed by the genius of great Russian military commanders who were to give forms to the expression of military art most appropriate to the needs of the Russian army.[119]

As an illustration Mikhnevich focused his attention on Skobelev, hero of the Russo-Turkish War of 1877/78, who had led the only successful attacks against fortified Turkish entrenchments at Shipka-Sheinovo and during the siege of Plevna. This was a situation where the Russian expression of military art was far superior to that of the West in that the tactical forms Skobelev had adopted had set the standard for European infantry regulations of that era. It was also to provide an excellent example of the importance of the high moral quality of Russian troops who, during this engagement, were capable of continuing the attack while suffering up to 40 percent casualties—a feat unattainable by the troops of any other European nation. This in turn proved that Russian military art was at a higher level compared to that of its neighbors, a situation which would allow Russian troops to handle twice as many foreign troops. Skobelev's famous storming of the Turkish camp at Shipka-Sheinovo was not only to be seen as the transcendence of Russian military art over European forms but as the sublime expression of the moral/cultural continuum between tactics and the historical calling of the nation: "The high quality of the soldier promotes more decisive action in battle; strong tactics allows a more ambitious strategy, which in turn allows the politician to pursue wider historical goals."[120] This sacrosanct continuum was the source of national expression in military art, distinct from the universal, eternal principles of military art. If defied by the attempt to borrow forms of military art from others, failure would inevitably result.[121]

However, the importance of more subjective elements of military art was not to deny the importance of studying the material factors of war because, as Yazykov, Miliutin, Neelov, Astaf'ev, and Leer had asserted, the essence of military art did not change, only the forms of its application—in this case significantly determined by technological innovations. In the monograph *Influence of New Technological Innovations on the Tactics of Troops* (*Vliianie noveishikh tekhnicheskikh ito bretenii na taktiku voisk*), Mikhnevich was to devote his undivided attention to this second element of military art. In order to approach the complexity of the interaction of technology with material forms of tactics, he established a classification of the significance of these new capabilities: (1)"perfection of the means of victory" (magazine weapons, smokeless powder, field mortars, quick-firing weapons, cartridge ammunition, self-loading weapons, grenades, etc.); (2) "perfection of the means of moving troops" (railroads, steam-powered navy, and bicycles);[122] (3) "perfection of the means of communication" (telegraph, telephones, signals, observation balloons, etc.); and (4) "inventions with the goal of improving the troops' way of life" (canned foods, sanitation, etc.).[123]

Turning to the study of the interaction of technology and tactical forms of military art, Mikhnevich again focused on Skobelev's single battle at Shipka-Sheinovo as the epitome of modern warfare revealing the future forms of infantry tactics. Perhaps the cardinal lesson to be learned from this battle was that any significant improvement of artillery was to alter the balance between the offense and the defense.[124] One such change was brought about by smokeless powder which no longer provided a screen with which the attacker could approach the opponent undetected, the result being that the organization of the attacking force was more dependent on the configuration of terrain for cover.

Another development was the quantitative and qualitative increase of quick-firing and long-range artillery which, when coupled with the advantages that indirect fire positions allowed the commander, could provide an increased economy of force to the defensive. This atomizing quality of technology was also apparent in the resulting forms of the offense which further emphasized the regard for protective cover in terrain and the need to decrease the size of attacking formations as well as increase the space dividing them—a resulting complexity in the offensive battle order which required the regularization of the order of attack.[125]

All of these observations were to point to the need to reexamine the field regulations of 1881 which Leer had compiled with the hope of bringing them up to date with an analysis of the possible forms of future warfare. Although Leer's work had improved upon methods standardized by previous field regulations, he had not benefited from extensive discussion of all the important implications for tactics; thus tactical training remained imperfect.[126] The previously insufficient analysis of Skobelev's campaigns in the Russo-Turkish War in conjunction with the continual rearmament of potential West European opponents was to necessitate a recompilation of these field regulations according to the following principles:

> 1.) recognize the independence of platoons and companies when under heavy enemy fire in the front lines; 2.) diminish the size of a company's sector in battle by one half while depth may be retained as before; 3.) allow the execution of attacks (bayonet charges) by leaps and bounds as they were done in the majority of cases by our troops in the last war; 4.) today it is not uncommon to reserve the attack for those cases when the preparation of its fire develops to such an extent that the opponent begins to abandon his position before the movement of our troops into a bayonet charge.[127]

The thread of argument weaving through these two works was to lead to a conclusion which had long eluded Russian military theorists: the direct translation of highly abstract and subjective characteristics of Russian military theory into concrete forms of military operations. This unified view of military art was the genesis of a whole new concern of early Soviet military theory: the formation of a unified military doctrine.[128]

The shift in power relations in Europe brought about by Prussian military victory over the French in 1870/71 profoundly influenced the direction of Russian military strategic priorities up to the First World War. From the outset of this new era War Minister Miliutin and his colleague from the Nicholas Academy, Obruchev, reflected their personal and institutional biases in the formulation of strategic plans. In favor of a stronger French Republic to offset any threat the new Germany in alliance with Austro-Hungary might have against the aspirations of the Russian Empire, they had begun a systematic compilation of strategic plans for a potential military conflict with these neighboring powers as early as 1866.

Although this liberal group of military reformers were often at odds over most issues of military affairs with the more reactionary members of the imperial family and senior military commanders such as Grand Duke Nicholai Nicholaevich (who during the Russo-Turkish War had refused to allow Obruchev to serve as chief of staff to the heir apparent), on the issue of cordial relations with representatives of the French army there was no disagreement.[129] Whereas the Grand Duke and his colleagues maintained purely personal contacts with French generals and statesmen, Miliutin and Obruchev conducted their relations at a highly official and confidential level through military institutions of higher education and research, namely, the main staff, military district staffs, and the Nicholas Academy.[130]

Beginning in 1872 this relationship manifested itself in the increased access of the French Ambassador, General Le Flô, and his military attaché, Colonel Gaillard, to information concerning the material strength, organization, and training of the Russian army. Prior to the Russo-Turkish War, Gaillard had been given the opportunity to observe Russian mobilization procedures and upon commencement of hostilities with Turkey had accompanied the Grand Duke as part of his staff in the theater of war.[131]

After the conclusion of the Russo-Turkish War, the high command of the Russian army began to readdress plans for hostilities along its western frontier. Intent on finding some way to neutralize Austrian military power, a special conference was held at the Nicholas Academy in May 1886 to determine what forces could be mobilized in the event of a war on the western frontier and how these forces could be combined using the existing railroad network.[132] The conclusion reached at this conference was that over the past four years a sufficient amount of new railroads had been built to speed up the mobilization of active troops from three months to forty days.[133]

Within the same month a delegation of Russian military observers—headed by Major-General Feldman, secretary of the technical committee of the general staff, and accompanied by Puzyrevskii—was sent to France to observe the grand maneuvers of the French army. In reporting their departure to his superior, the French military attaché remarked that "there is no doubt that the impressions which he [Feldman, and Puzyrevskii as well] gains during his mission in France will have a significant influence on the Chief of the General Staff's [Obruchev's] opinion [of the French army]."[134] Later, in a letter to his friend and former chief, Obruchev indicated the direction for future strategic planning while implicitly acknowledging the role of France in the new strategic calculus:

To me it seems beyond question that for Russia there are only two basic historical questions for the sake of which it would be worth shedding Russian blood. These are the Polish [Galician] question and that of the Bosphorus. Whether one wishes it or not, these questions will have to be decided by war; and this, precisely, is the war for which we have to prepare.[135]

When relations with Austria and Germany took a further turn for the worse during the threat of war in the winter of 1887/88, the Russian main staff responded by drawing up the Plan of Concentration No. 1 which was completed in mid-February 1888. This was the first war plan which took into account the possibility of a Russo-Franco military coalition against the combined forces of Austro-Hungary and Germany. Due to the inadequacies of the rail network which delayed the mobilization of Russian troops to their respective assembly points and eventual concentration in the theater of war, Obruchev was limited to two strategic options. One was to concentrate forces well behind the frontier in order to attack the enemy with the full force of Russian military power. This scenario was the most acceptable to the majority of the general staff and was most likely to succeed if Russia was forced to fight Germany and Austria by itself. If France were also involved, however, it would give Germany two months to neutralize France before it would have to shift its forces to the east.

The second scenario, accepted as the official plan of concentration if Russia was joined by France against allied German and Austrian military forces, proposed that this long period of mobilization be avoided by committing the most quickly transportable of Russian contingents to present an immediate threat while the remainder of the Russian army completed its full mobilization. Although it was not expected to achieve the same results as the fully mobilized forces described in the first scenario, this option was seen to provide more support for French forces beleaguered by a German invasion. This was the first official suggestion for French and Russian cooperation in operational affairs. Obruchev was anxious to test the waters of French military politics via the French military attaché in St. Petersburg, Captain Louis Moulin, for reassurance of such a coalition.[136]

At this time the majority of the Russian general staff felt that Russian forces would strike an offensive blow against Austria at the beginning of hostilities. The plan proposed to sweep through Galacia, occupy the Carpathian Mountains and the valley of the Morava River up to its confluence with the Oder River, then march straight for Berlin through Prussian Silesia, assuming that the German army would still be occupied in a war with France.[137]

The role of Gurko's Polish army, committed to the defense of the Polish Salient against a possible German and Austrian offensive, posed difficult problems for strategic planning which defied simple resolution. Since the tendency of Russian strategic planning under Obruchev was toward an offensive against Austrian Galicia, Gurko's Polish army was constantly left with too few troops to reasonably expect to halt a German advance on the left bank of the Visla and Narev rivers, and had to continually worry about Austrian efforts from the south to cut off its retreat. When the Reinsurance Treaty between Germany and Russia failed to be renewed upon its expiration in 1890, the final obstacle in Franco-Russian diplomatic and military relations was finally removed and precipitated the signing of the Franco-Russian Military Convention in August 1891.

In spite of the fact that stipulations made in the treaty called for the close coordination of strategic planning between the two general staffs, Obruchev managed to maintain his freedom of action by remaining relatively inaccessible to representatives of the French general staff. Instead, they were referred to the new chief of staff at the Warsaw military district, Puzyrevskii, who had recently been transferred from his post at the Nicholas Academy.[138] In these briefings Puzyrevskii gave to the French military attaché, Moulin, extensive information concerning the various strategic options available to the southwestern army and, more specifically, the Polish army.[139]

Puzyrevskii remained highly skeptical about any offensive role to be played by Gurko's Polish army and assumed that he would have to plan for a German offensive from East Prussia. Drawing conclusions from the statistical analysis of various scenarios given to the newly created statistical section of the Warsaw military district, he counted on a German attack on the junction of the Bug and Narev rivers near the center of the Bielostok/Brest-Litovsk railroad line down which the Polish army would be forced to retreat.[140] Nor was the Polish army to expect any relief from the south. The principal axis of the Austrian offensive was not expected to be diverted toward Kiev but to join in attempting to cut off the Polish army's retreat to Bielostok. On this subject Puzyrevskii was anxious to find out what calculations the French had made in determining how fast the Austrians would march to the Bug River.[141]

By 1902 these strategic options were supplemented by a hypothesis for an isolated offensive by the Polish army from the left bank of the Narev River directed toward Prussia, but planners at the Warsaw military district were forced to scrap it because it could not make use of any railroad support, did not correspond to the probable concentration of German forces or disrupt this concentration, and did not take into consideration the threat of the Austrian army from the south.[142] Indeed, the French military observer, Captain Mahon, who was allowed to review these different strategic hypotheses, concluded that the planning effort was marked by indifference and a harmful amount of eclecticism, much the same as F. Gershal'man, assigned to the staff of the Warsaw military district in 1902, had found it:

In matters on the defense of the western frontier nothing was definite, nothing finished. We dissipated ourselves in a mass of frequently contradictory assumptions, touching on everything, stopping for nothing, and eliciting in ourselves only complete distrust for our decisions which changed constantly. . . . Under these conditions we were completely unprepared for war on the Western Front.[143]

This lack of rationalization in strategic planning for the defense requirement of a nation facing an ever-increasing military potential from its western neighbors was both a function of the increasing operational responsibilities of the planning bodies and their growing administrative obsolescence. By the mid-1890s the implications of the *Regulations on the Field Administration of Forces in Wartime,* published in 1889 and introducing the need for army groups to coordinate operations over the extensive western frontier, had become apparent to strategic planners at the main staff. While regional strategic planning needs were met by the integration of new statistical sections within the staff functions of the frontier military districts, the chief of staff, Obruchev, proposed to

reorganize the Main Staff in such a way as to give it more participation in strategic matters such as the distribution of troops in the Theater of Military Operations, the composition of the army, the creation of war plans for the initial phase of the war, preparations of field administrations from frontier Military District staffs and the collection and analysis of intelligence on the composition of enemy armies.[144]

To investigate the feasibility of this reorganization, a special commission presided over by Gurko was established in 1894. By 1896 it had come to the conclusion that the war ministry be broken up by its administrative, operational, and planning functions into three new ministries enjoying direct access to the tsar: the war ministry under General Lobko (former professor of military administration at the Nicholas Academy and responsible for the compilation of the new *Regulations*), the inspector-general of the army under Dragomirov, and the chief of the main staff under Obruchev.[145] Needless to say, these recommendations suited no one and all efforts to reorganize the central administration fell by the way when War Minister Vannovskii and his chief of staff resigned abruptly at the end of 1897. Obruchev was still keen to address the problem and in February 1898 resubmitted a proposal to the new war minister, Kuropatkin (a former student and colleague), recommending specialization of main staff duties while granting the chief of the main staff access to the tsar upon the war minister's absence.[146]

By March 1900 this proposal was accepted by both the war council and the tsar with the immediate creation of two new administrations, the 1st and 2nd quartermaster general followed by three more administrations in 1903.[147] Unfortunately, these were insufficient reforms to tackle the required systematization of the Russian army's strategic planning process. Frontier military districts were still solely responsible for their own mobilization schedules and plans of concentration in their specialized theater of military operations with the provisional field staffs responsible for actual deployment and initial march-maneuver plans.[148] The absence of regional and hierarchical coordination in the strategic planning process was partially ameliorated by the convening of occasional strategic planning conferences in St. Petersburg composed of the frontier military district staffs under the presidency of Obruchev.[149] Ultimately, even these ad hoc consultations at the center were insufficient to provide effective and efficient purpose to the planning process, inadequacies which were to come to the fore in the approaching war in the Far East.

NOTES

1. P. A. Zaionchkovskii, *Voennye reformy 1860–1870 godov v Rossii,* (Moscow: Izdatel'stvo Moskovskogo Universiteta, 1952), 293.
2. P. A. Zaionchkovskii, *Samoderzhavie i russkaia armiia na rubezhe XIX-XX stoletti, 1881–1903* (Moscow: Izdatel'stvo 'Mysl', 1973), 92.
3. Ibid., 59.
4. Ibid., 62. Although Obruchev was not appreciated by Alexander III, his capabilities as chief of the main staff, dominant role in the formulation of strategic plans, and his supervision of the military-scientific work of the general staff were to win him widespread praise from such disparate figures as S. Iu. Witte, D. A. Miliutin, A. N. Kuropatkin, F. K. Gershel'man, and P. S. Vannovskii.
5. Ibid., 93. In choosing a chairman for the commission which was to examine possible reorganization of Miliutin's reforms in administration, Vannovskii appointed P. E. Kotseby instead of Obruchev, much to the relief of conservative groups who recalled Obruchev's revolutionary tendencies as a cadet in the 1850s. Obruchev was still appointed as a member of the commission and was subsequently able to defuse any serious alteration of Miliutin's reforms.
6. Ibid., 96–97.
7. Ibid., 97–99.
8. Ibid., 97. At this meeting Dragomirov was to lend his support for maintaining the consolidated war ministry as it had been created under Miliutin.
9. Ibid., 102.
10. G. V. Malanchuk, "The Training of an Army: M. I. Dragomirov and the Imperial Russian Army, 1860–1905" (M.A. thesis, Miami University, 1978), 41.
11. F. A. Maksheev, "Neskol'ko slov o Dragomirovskom i Leerovskom periodakh nachal'stvovaniia akademiei," *Voennyi sbornik* 12 (1907): 234–35. See also A. Avdeev, "Voenno-istoricheskie issledovaniia v akademii general'nogo shtaba russkoi armii," *Voenno-istoricheskyi zhurnal* 12 (1987): 78.
12. Maksheev, "Neskol'ko slov o Dragomirovskom i Leerovskon periodakh nachal'stvovaniia akademiei," 236. See also P. A. Zhilin, pod. red., *Russkaia voennaia mysl' konets XIX–nachalo XX v.* (Moscow: Izdatel'stvo 'Nauk', 1982), 47.
13. Zaionchkovskii, *Samoderzhavie i russkaia armiia na rubezhe XIX–XX stoletii, 1881–1903,* 176.
14. Ibid., 179. A. N. Kuropatkin also shared these views; see his memoirs, *The Russian Army and the Japanese War: Being Historical and Critical Comments on the Military Policy and Power of Russia and on the Campaign in the Far East* (London, 1909), 1:101.

15. N. P. Glinoetskii, *Istoricheskii ocherk Nikolaevskoi akademii, general'nogo shtaba,* (St. Petersburg, 1882), 318–23.

16. Ibid., 323–24. Kuropatkin was evidently one of Obruchev's and Glinoetskii's protégés. In 1879 he published a major study entitled *Kashgariia: Istoriko-geograficheskii ocherk' strany, eiia voennyi sily, promyshlennost' i torgovlia,* and also contributed to *Obzor voin Rossii ot Petra Velikago do nashikh dnei.*

17. Zhilin, *Russkaia voennaia mysl' konets XIX–nachalo XX v,* 36. See also G. P. Meshcheriakov, *Russkaia voennaia mysl' v XIX v* (Moscow: Izdatel'stvo 'Nauka', 1973), 254.

18. Zaionchkovskii, *Samoderzhavie i russkaia armiia na rubezhe XIX–XX stoletii, 1881–1903,* 103–7. It is very possible that this commission work on the field regulations done by Leer and in turn by Puzyrevskii was to provide the main fuel behind their disagreements over the experience of the Russo-Turkish War on the nature of military art, and consequently on the reform of field command hierarchy and structure and function which was evident in their later theoretical works.

19. Carton 7N 1484, Captain P. Dreyfus, "Étude sur l'armée russe. Le Général Dragomiroff" (Perpignan, May 26, 1896), 16–17.

20. Malanchuk, "Training of an Army," 45.

21. Carton, 7N 1484, Dreyfus, "Étude sur l'armée russe," 17–18.

22. B. W. Menning, "Bayonets before Bullets: The Imperial Russian Army, 1861–1914" (unpublished manuscript, 1989), 236.

23. Malanchuk, "Training of an Army," 44.

24. Carton 7N 1484, Dreyfus, "Etude sur l'armée russe," 19.

25. Carton 7N 1470, Captain Moulin, "Les manoeuvres 'de bataillon' proposées par le Général Dragomiroff," (Oct. 25, 1886).

26. Menning, "Bayonets before Bullets," 261–62.

27. Carton 7N 1471, Moulin, "Opinion du Général Stroukoff sur les raids en Prusse Orientale" (Oct. 10, 1889); Carton 7N 1472, Moulin, "Rapport sur les manoeuvres de l'armée russe en Volkynie" (Nov. 15, 1890); Carton 7N 1472, Moulin, "Sur la revision du règlement d'infanterie" (Dec. 8, 1894).

28. Carton 7N 1469, Lt.-Colonel de Sermet, "Échanges de documents proposés par le Général Dragomiroff entre les deux écoles superiéures de guerre" (Nov. 9, 1884; Feb. 10, 1885).

29. Menning, "Bayonets before Bullets," 24.

30. Carton 7N 1471, Moulin, "Opinion du Général Leer sur l'état de preparation à la guerre de l'armée russe" (June 6, 1887). Moulin, assistant military attaché to the French embassy in St. Petersburg at this time, characterized Leer in the following terms: "It is enough . . . to have read his strategy and above all his conferences on the war of 1870 to recognize in him a spirit almost as systematic as that of Pfuhl, the famous strategen of the Campaign of 1812."

31. *Konspekt strategii,* sostavlen ofitserami starshogo klassa Nikolaevskoi akademii general'nogo shtaba (St. Petersburg, 1887), 1.

32. Ibid., ii.

33. Ibid., 4.

34. Ibid., 10. The compilers of the *Konspekt strategii* provided a schematic diagram outlining the recommendations of theory for paths of attack and maintenance of communication in the context of the organization of the rear similar to one provided by Leer in his *Opyt kritiko-istoricheskogo izsledovaniia zakonov iskusstva vedeniia voiny.*

35. Ibid., 11–12.

36. Ibid., 3.

37. L. Leonidov, "Voennyi istornik A. K. Puzyrevskii," *Voenno-istoricheskyi zhurnal* 1 (1969): 97. Puzyrevskii's extensive experience in the Russo-Turkish War of 1877/78 was to be reflected in his first serious published works: "Otryvochnye zametki iz opyta minuvshei kampanii," *Voennyi sbornik* 7 (1878); and *Vospominaniia ofitsera general'nogo shtaba o voine 1877–1878 godov v Evropeishoi Turtsii* (St. Petersburg, 1879), which was based on a sixteen-part series entitled "Pisem s teatra voiny," published in *Russkii invalide* in 1877.

38. Ibid., 92.

39. Avdeev, "Voenno-istoricheskie issledovaniia v akademii general'nogo shtaba russkoi armii," 78. See also L. G. Beskrovnyi, *Ocherki voennoi istoriografii Rossii* (Moscow: Izdatel'stvo Akademii Nauk SSSR, 1962), 229.

40. Leonidov, "Voennyi istornik A. K. Puzyrevskii," 95.

41. Beskrovnyi, *Ocherki voennoi istoriograffi Rossii,* 219.

42. Ibid., 224–26. His work in compiling new archives and publishing new collections of primary sources was epitomized by his *Opis' del' sekretnago pobytiia moskovskaog otdeleniia obshchago arhkiva glavnago shtaba (47-iia).* Maslovskii was to emphasize this in the comment that "the success and virtue of research, of course, depends not only on the quantity but also on the quality of the collected material and on its critical treatment" (cited in *Zapiski po istorii voennogo iskusstva v Rossii,* issue 1, 3–4).

43. Ibid., 230.

44. Leonidov, "Voennyi istornik A. K. Puzyrevskii," 95.

45. Ibid.

46. Ibid.

47. Puzyrevskii relied on a mixture of French and German bibliographic sources as well as the work of Stankevich and the unpublished work of his colleague Sukhotin. He provided an extensive bibliography and footnotes, the first to be seen in works of this era. This was a continuation of his earlier work, *Istoriia voennogo iskusstva v srednie veka* (*History of Military Art in the Middle Ages*).

48. Ibid., 3.

49. Leonidov, "Voennyi istornik A. K. Puzyrevskii," 97.

50. Beskrovnyi, *Ocherki voennoi istoriografii Rossii,* 220 (cited in D. F. Maslovskii, "Iz istorii voennogo iskusstva v Rossii. Opyt kriticheskogo razbora pokhoda Dmitriia Donskogo 1380 g. do Kulikovskii bitvy vkliuchitel'no," *Voennyi sbornik* 8–9 (1881): 208.

51. Zhilin, *Russkaia voennaia mysl' knoets XIX–nachalo XX v,* 161.

52. Beskrovnyi, *Ocherki voennoi istoriografii Rossii,* 221.

53. Avdeev, "Voenno-istoricheskie issledovaniia v akademii general'nogo shtaba russkoi armii," 79.

54. G. A. Leer, "Sopostavlenie neskol'kikh evoliutsii v strategicheskom iskusstv'," *Voennyi sbornik* (1893): 68–69.

55. Ibid., 90.

56. P. Geisman, *Kratkii kurs istorii voennogo iskusstva v srednie i novye veka* (St. Petersburg, 1893–1896), 2: "military history makes up a section of the theory of military art, enabling its practical study and even attaining the ability to help at certain times all possible means for attaining the goals of the war, i.e., up to a certain point it fills in the gaps of personal experience."

57. Ibid., iii.

58. V. A. D'iakov, "O razvitii russkoi voenno-istoricheskoi mysli v poslednei chetverti XIX veka," *Voenno-istoricheskyi zhurnal* 5 (1959): 66.

59. Geisman, *Kratkii kurs istorii voennogo iskusstva v srednie i novye veka,* i. See also D'iakov, "O razvitii russkoi voenno-istoricheskoi mysli v poslednei chetverti XIX veka," 65.

60. Ibid., 3.

61. Meshcheriakov, *Russkaia voennaia mysl'v XIX v,* 264.

62. Beskrovnyi, *Ocherki voennoi istoriografii Rossii,* 210.

63. P. A. Geisman, *Opyt issledovaniia taktiki massovykh armii* (St. Petersburg, 1894), 274.

64. Ibid., iv, 277.

65. Ibid., 235.

66. Ibid., 237.

67. Goerlitz illustrated these concepts by citing Moltke's critique of the old style of command: "The advantage which a commander thinks he can attain through continued personal intervention is largely illusory. By engaging in it he assumes a task which really belongs to others, whereas effectiveness he thus destroys. He also multiplies his own tasks to a point where he can no longer fulfill the whole of them" (W. Goerlitz, *The History of the German General Staff, 1657–1945* (New York: Praeger, 1953), 76.

68. Geisman, *Opyt issledovaniia taktiki massovykh armii,* 238.

69. E. I. Martynov, *Strategiia v epokhu Napoleona I i v nashe vremiia* (St. Petersburg, 1894), i, 269.

70. A. A. Strokov, *Istoriia voennaia iskusstva* (Moscow: Voenizdat, 1965), 600.

71. Martynov, *Strategiia v epokhu Napoleona I i v nashe vremiia,* 8–9.

72. Ibid., 3.

73. Ibid., 290.

74. Ibid., 293.

75. Ibid., 29, 268.

76. Ibid., 275–76.

77. Ibid., 278.

78. Ibid., 279–80.

79. Ibid., 284, 300. Martynov was to capture this evolution of strategic styles with the archetypal Leerian statement that "cavalry is the main proponent of the moral element in war, artillery embodies in itself the idea of material strength, the idea of technology. The uninterrupted reduction of relative numbers of the first and the rise in the second serves as evidence that the significance of the moral element in war falls more and more to the superiority of technology."

80. Ibid., 301.

81. A. N. Petrov, "Zadachi sovremennoi strategii," *Voennyi sbornik* 5 (1894): 41.

82. Ibid., 43.

83. Ibid.

84. Ibid., 35.

85. Ibid., 36–37.

86. Ibid., 40.

87. A. N. Petrov, *K voprosam strategii* (St. Petersburg, 1898), 351.

88. A. N. Petrov, *Vlianie turetskikh voin s poloviny proshlogo stoletiia na razvitie russkogo voennogo iskusstva* (St. Petersburg, 1893/94), 338.

89. Petrov, *K voprosam strategii,* 352.

90. Petrov, *Vlianie turetskikh voin s poloviny proshlogo stoletiia na razvitie russkogo voennogo iskusstva*, 338.

91. In his *Nation in Arms*, originally published in 1883, von der Goltz was the first to glimpse the full enormity of modern warfare brought about by universal male conscription, the strategic implications of railroads, and technological advances in armament.

92. Petrov, *K voprosam strategii*, 357.

93. Strokov, *Istoriia voennaia iskusstva*, 639.

94. Menning, "Bayonets before Bullets," 234.

95. Petrov, *K voprosam strategii*, 354. Both Petrov and Geisman shared this technique in discussing the components of military power. Even this was cause for disagreement with Martynov regardless of the argument being made. In the only marginalia made in his personal copy of *An Attempt at a Theory of Tactics for Mass Armies*, Geisman reacted strongly to Martynov's cynical comments that his equation was "a formula for victory" (author's copy, 224).

96. Ibid., 361.

97. Petrov, "Zadachi sovremennoi strategii," 46–48.

98. Ibid., 49.

99. Petrov, *K voprosam strategii*, 362.

100. Petrov, "Zadachi sovremennoi strategii," 61.

101. Carton 7N 1480, M. Bouard, French council general in Warsaw to the minister of foreign affairs, "La guerre future et ses consequences économiques" (Apr. 24, 1893).

102. I. S. Bliokh, *Obshchie vyvody iz sochineniia "Budushchaia voina v tekhnicheskom, politicheskon i ekonomicheskom otnosheniiakh"* (St. Petersburg, 1897), 46.

103. Ibid., 48.

104. Ibid., 163.

105. Ibid.

106. M. I. Dragomirov, "Otkrytoe pis'mo G. Bliokhu," *Razvedchik* 316 (1898).

107. A. K. Puzyrevskii, "Budushchaia voina v tendentsioznom izobrazhenii," *Razvedchik* 410 (1898).

108. A. Gulevich, "Voina i narodnoe khoziaistvo," *Voennyi sbornik* (1898): 73.

109. Ibid., 60.

110. Ibid., 63.

111. Ibid., 66.

112. Ibid.

113. Ibid., 73. Leer had begun to analyze this phenomenon starting in the 1887 edition of *Strategiia* and developing it in each subsequent edition up to the final edition in 1898.

114. Carton 7N 1475, French ambassador in Russia to the minister of foreign affairs, "La retraite des généraux Vannovsky et Obroutcheff" (Jan. 19, 1898).

115. Carton 7N 1475, Moulin, "Sur le Général Kouropatkine" (Oct. 11, 1898). This observation was made by S. Witte after the fiasco over the Muraviev Circular calling for the Hague Peace Conference.

116. Carton 7N 1475, Moulin, "Nouvelles nominations dans l'armée" (Feb. 10, 1898); "Changements dans le haut personnel de l'armée russe" (Feb. 25, 1898); and "Nouvelles nominations de gouverneurs généraux" (Apr. 10, 1898). When Puzyrevskii learned that he had been passed over for the appointment of chief of staff, he threatened to resign his post at the Warsaw military district. After overtures were made to appoint him to the governorship of Turkestan and the Transcaucasus, he eventually agreed to remain at his old post.

117. Kuropatkin, *Russian Army and the Japanese War,* 1: 39.

118. N. P. Mikhnevich, *Osnovy russkogo voennogo iskusstva. Sravnitel'nyi ocherk sostoianiia voennogo iskusstva v Rossii i Zapadnoi Evrope v vashneishiia istoricheskiia epokhi* (St. Petersburg, 1898), 4.

119. Ibid., 6.

120. Ibid., 175.

121. Ibid., 170.

122. In lectures given at the Nicholas Academy in late 1896, A. von Vendrikh was to call for the development of a special branch of military-scientific knowledge, hitherto neglected, on the significance of railroads in military relations to be conducted in a special institution and to draw upon the expertise of civilians as well as foreign military specialists. This new branch of military science was to be based on Leer's criticohistorical analysis and his work on tactics and strategy, and was to adapt to the territorial military administrative system. See A. A. von Vendrikh, *Zavisimost' budushchikh voin ot blagoustroistva voennykh soobshchenii: soobshchenie Nikolaevskii akademii general'nogo shtaba* (St. Petersburg, 1889), 2–4, 437–38, 439–40.

123. N. P. Mikhnevich, *Vliianie noveishikh tekhnicheskikh izobretenii na taktiku voisk* (St. Petersburg, 1898), 416.

124. Ibid., 425.

125. Ibid., 423–24. Impressed by the significance of this battle, some French military theorists supported the argument for extended order in tactics and reliance on the development of fire tactics as expressed by the French service regulations of 1875 for the formulation of new regulations in 1884. See J. C. Arnold, "French Tactical Doctrine, 1870–1914," *Military Affairs* (Apr. 1978) for more discussion of mutual French/Russian exchange of tactical concepts up to the First World War.

126. Ibid., 417.

127. Ibid., 440.

128. In 1898, when Kuropatkin was appointed to replace Vannovskii as war minister, he instigated a review of military policy to be conducted by Rediger and other instructors at the Nicholas Academy. He eventually submitted this review to the tsar as a report in 1900. This report, to which reference is restricted to quotes used by Kuropatkin in his later apologist study of the Russo-Japanese War, is perhaps the greatest geostrategic study of all time. That it was a collaborative analysis which may have used the unified military-scientific method and the systems-analysis approach of Gulevich and Rediger suggests a very interesting and powerful direction in Russian military science at this time (Kuropatkin, *Russian Army and the Japanese War,* xxiv–xxv; the first two hundred pages relate to this assessment and the report).

129. G. F. Kennan, *The Decline of Bismarck's European Order: Franco-Russian Relations, 1875–1890* (Princeton: Princeton University Press, 1979), 52.

130. Ibid., 52.

131. Ibid., 207. See also F. Wellesley, *Recollections of a Soldier-Diplomat* (London: Hutchinson, n.d.), 129.

132. Carton 7N 1470, Moulin, "Forces russes sur l'éventuels theatre d'opérations" (May 1, 1886).

133. Carton 7N 1470, Moulin, "Plan de concentration en cas de guerre conte l'Allemagne et l'Autriche" (June 12, 1886).

134. Carton 7N 1470, Moulin, June 19, 1886.

135. Kennan, *Decline of Bismarck's European Order,* 152.

136. Carton 7N 1471, Moulin, "Envoi du plan de concentration no. 1 pour 1889" (Oct. 10, 1889).

137. Carton 7N 1471, Moulin, "Opinion du Général Stroukoff sur les raids en Prusse Orientale" (Oct. 10, 1889).

138. Menning, "Bayonets before Bullets," 193. See also Leonidov, "Voennyi istornik A. K. Puzyrevskii," 91; and Carton 7N 1472, Moulin, "Rapport sur les manoeuvres de l'armée russe en Volkynie" (Nov. 15, 1890).

139. Carton 7N 1473, Moulin, "Conversation avec un général russe au sujet du plan d'opération de l'armée de Gourko" (Dec. 22, 1893).

140. Carton 7N 1485, Captain Mahon, "Rapport sur un séjour à Varsovie," (Oct. 1902). Statistical sections became an essential part of the military district staff work after the general reforms of 1893. These sections also regulated the conduct of staff war gaming during the winter months and staff rides in the summer in order to evaluate the configuration of the strategic plans.

One of the more interesting responsibilities of the statistical section at the Warsaw military district was the continual assessment of the use of frontier brigades to screen the full concentration of the Russian army from an enemy offensive at the beginning of a war. Operating between the Narev River and the border, these special troops, in conjunction with independent cavalry and partisans and capable of falling back on advance infantry positions *(spitztruppen),* were highly responsive to the strategic goals and realities presented by the approaching hostilities. These specially trained troops formed the nucleus of a new type of mobile, active-defensive warfare which was to manifest itself in the Russo-Japanese War of 1904/5 and again during the Civil War in the early years of the Soviet state.

141. Carton 7N 1473, Moulin, "Conversation avec un général russe au sujet du plan d'opération de l'armée de Gourko" (Dec. 22, 1893).

142. Carton 7N 1485, Captain Mahon, "Rapport sur un séjour à Varsovie. Section III, Rapports particulars avec les officiers de l'entourage du Général Pouzerevski" (Oct. 1902).

143. Menning, "Bayonets before Bullets," 196.

144. L. G. Beskrovnyi, *Russkaia armiia i flot v XIX veke. Voenno-ekonomicheskii potentsial Rossii* (Moscow: Izdatel'stvo 'Nauka', 1973), 206.

145. Carton 7N 1474, Moulin, "Reúnion d'une commission presidée par Gourko" (Jan. 30, 1896).

146. J. Erickson, "The General Staff: Theory and Practice from the Tsarist to the Soviet Regime," in *National Security Policy: The Decision-Making Process,* edited by A. L. Pfaltsgraff, Jr., and U. Ra'anan (Medford, Mass.: Archon, 1984), 20–21.

147. Ibid., 21; and L. G. Beskrovnyi, *Armiia i flot Rossii v nachale XX v: Ocherki Voenno-ekonomicheskogo potentsiala* (Moscow: Izdatel'stvo 'Nauka', 1986), 48–49. The 1st quartermaster general consisted of the 6th and 7th sections, the former responsible for the military geography and statistics of Russia, and the latter responsible for collecting foreign statistics and compiling civilian and military information on foreign governments as well as maintaining the archives and library at the main staff. The 2nd quartermaster general consisted of the military statistical department and the operational department. From the Hoover Institution of War, Revolution, and Peace, E. A. Nikol'skii Collection, "Sluzhba v Glavnom shtabie i Glavnom upravlenii general'nogo shtaba (1903–1908)," 3.

148. Menning, "Bayonets before Bullets," 195.

149. Carton 7N 1472, Moulin, "Réunion des chefs d'État-Major de Varsovie, Vilna et Kief à Petersbourg" (Feb. 1, 1895).

5

Modernization: The Impact of Technology on Military Science and the Search for a Unified Military Doctrine, 1904–1914

In early February 1904 the Japanese fleet under the command of Admiral Togo Heihachiro conducted a surprise attack on the southern Manchurian stronghold of the Russian Pacific fleet at Port Arthur, precipitating war with the colossal European power which had demonstrated expansionist military and economic policies in the Manchurian and Korean peninsulas from 1897. From the outset of the hostilities the Russian defense was hampered by disagreement between the recently appointed viceroy in the Far East, Admiral E. Alekseev, who insisted on the active defense of Port Arthur, and members of the high command in St. Petersburg, who recommended the evacuation of southern Manchuria for consolidation of military forces in northern Manchuria.[1] The existing level of military preparation in the region was also insufficient for defense against a significant invasion. Although work on fortifications and transport of more troops from European Russia and western Siberia had been initiated the previous summer in response to increasing diplomatic tensions, defense of both Vladivostok and Port Arthur still suffered from numerical inferiority which would take at least six months to rectify.

During this time the former minister of war, Kuropatkin, established command of the hastily created Manchurian army at Liao-yang and deployed two detachments as forward defense screens to stall the Japanese advance. The first detachment to make contact with the Japanese army marching north up the Korean peninsula was the eastern detachment deployed east of the Ya-lou River on the border with Korea. Initially preparing to disrupt the landing and concentration of the Japanese army at Pin-yang, the Russian commander and his troops were called back behind the Ya-lou by Kuropatkin and instructed to avoid decisive action against superior numbers.[2] On May 1, the eastern detachment near the village of Tur-en-chen attempted to stop the Japanese crossing of the Ya-lou to the north bank. What occurred there was to underscore the subsequent development of the war: the Japanese infantry applied the tactical doctrine it had learned from the German tactician, Colonel Meckel in the 1880s and 1890s, to threaten the envelopment of the Russian entrenchments at Tur-en-chen from the north and south while it pinned the entrenchments with heavy, indirectly laid artillery fire. Obliged to dissipate his forces to counter this maneuver, the Russian commander's combat initiative was further compromised by a continual relay of conflicting instructions from Kuropatkin in Liao-yang and Alekseev in Harbin. He was eventually forced to give up his position and slowly retreat over the Liao-tung Mountains to Liao-yang.[3]

By mid-August the combined force of Marshal Oyama's main army with the 1st Japanese Army under the command of General Kuroki finally confronted Kuropatkin's assembled Manchurian army immediately south of Liao-yang. In his famous Declaration of the Offensive to his troops, Kuropatkin promised that the moment for the long-awaited offensive against the Japanese invaders had arrived, convinced that an advantageous correlation of forces had been achieved.[4]

Over the next six weeks these two enormous armies engaged in a series of battles which were to represent a turning point in the war and in military history. Reapplying the tactical formula of envelopment which had been successful at Tur-en-chen, the Japanese forced Kuropatkin to fall back along his rail supply lines. In an attempt to conduct a counteroffensive, Kuropatkin abandoned the chaos of his army headquarters to provide direct leadership at the front, as his mentor Skobelev had done at Plevna a quarter-century before, and lost control of the battle.[5] Later, at the battle of Sha-ho, this combination of inefficient command and conflicting strategic authority confounded Kuropatkin's continued attempts to cross over to the promised offensive, forcing him to retreat still further up his rail supply line to Mukhden.

Early in January 1905 Port Arthur at long last fell to the Japanese siege under General Nogi and the bulk of this victorious army joined forces with Oyama's armies facing the Russian defenses at Mukhden. Kuropatkin was finally free from conflicting authority over the Russian forces in the Far East and exercised command of the first army group in Russian military history, the armies of Generals Linievich, Kaulbars, and Grippenberg.[6] However, he now had to contend with mounting criticism in St. Petersburg, which sought his replacement by a more offensive-minded general.[7]

In February Oyama made a final attempt to crush the Russian forces with a reapplication of the enveloping maneuver along a front which extended over one hundred kilometers and which severely challenged Kuropatkin's coordinated leadership of the immense forces under his command.[8] Witnessing these extended fronts and increased violence of combat, a Russian (and later famous Soviet) military theorist noted the emergence of a new priority in modern war: "Oyama was in effect out to prove that battles are lost and won not only in the minds of generals, but also in the capacities of modern staff organization and

Russo-Japanese War of 1904-1905
Situation at the end of April 1904

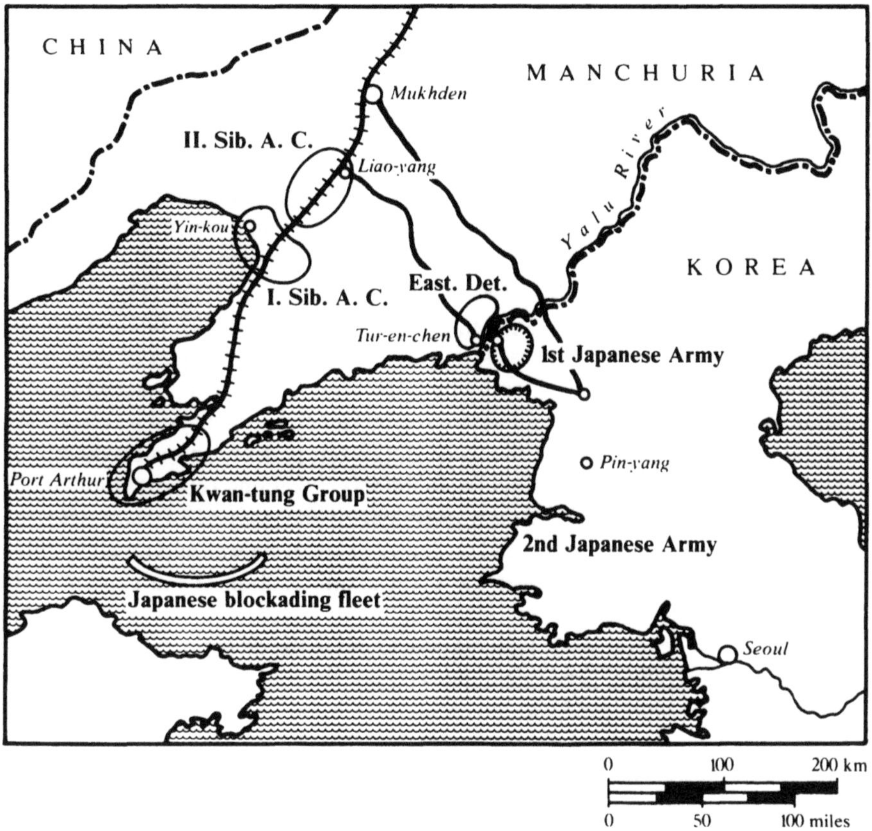

Source: Adapted from *The Russo-Japanese War,* prepared in the Historical Section of the German General Staff. Authorized translation by Karl von Donat, vol. 1. London: Hugh Rees, 1908.

subordinate commanders taxed with the necessity to react and adjust hurriedly to complex evolving situations."[9] Unable to gain the initiative after ten days of struggle along a front which extended over one hundred kilometers, Kuropatkin, under pressure from St. Petersburg, ceded command to Linievich. Under this new command the Manchurian army retreated up the railroad line to positions south of Harbin in what was to be the last defensive position before the sinking of the Russian navy in the Tsushima Straits and social unrest in European Russia were to bring an end to the war.

The lack of a coherent strategic plan on the part of the generals, a noticeable lack of command experience and knowledge on the part of the supreme commander and his subordinate officers, the eroded morale of the troops exhausted by successive defeats, and revolutionary agitation at home were seen as the main reasons for the Russian army's defeat. After the war the extensive debate on these issues and analysis of this most recent military experience within the institutions of the Russian army and navy were to influence the development of Russian military affairs, and especially military theory, up to the beginning of the First World War.

The confusion of command authority between Kuropatkin and Alekseev during the hostilities with Japan was a manifestation of urgently needed reform of the war ministry proposed by observant members of the Russian high command in the early 1890s. The first active step toward reform came when Sakharov, the newly appointed war minister, presented two proposals for the creation of a main administration of the general staff to the tsar in September 1904. The first proposal, drawn up by General Palitsyn, was based on the model of the German general staff in which the general staff was an independent organ of command separate from the war ministry and where the chief was directly subordinate to the tsar.[10] An alternative proposal submitted by General Engalichev was to create an organization responsible for the same duties but part of, and subordinate to, the war ministry with no special access to the tsar.[11] Gaining support from Grand Duke Nicholai Nicholaevich, future president of the council of state defense, Palitsyn's proposal was accepted and on June 21, 1905, he was appointed as chief of the general staff presiding over the main administration of the general staff inaugurated four days later.[12]

As chief of the general staff, Palitsyn was to guide the development of strategic plans and mobilization schedules, as well as the rest of the general staff's military-scientific work, and was to report to Nicholas II annually with a general account of his actions.[13] On December 7, 1905, in the midst of the demobilization of the Manchurian army, Palitsyn submitted a proposal on the administration of the general-quartermaster of the general staff which attended to questions of operational, organizational, and training significance to the Russian army. Accepting the premise that "the fighting preparation and operational abilities of the troops is directly dependent on the organizational arrangement of the armed forces, their stationing, training and education, as well as their mobilization," Palitsyn proposed that the planning responsibilities of the newly established institution would cover a wider spectrum of the nation's available resources as well as a more attentive analysis of developments in other countries.[14] With this proposal's approval by the military council in January 1906, Palitsyn had succeeded in creating an organization which, as E. I. Martynov was later to remark, would allow the "brain of the army to undergo a rationalizing treatment," and which was to prove to be very influential in the years of military reform up to the First World War.[15]

Spurred on by the recognition that the Russian army would do well to learn from its mistakes in the recent campaign, Palitsyn established under the military-historical department of the main administration of the general staff a special military-historical commission on the description of the Russo-Japanese War to be supervised by a special commission under the direct control of the chief of the general staff himself.[16] Avoiding the mistakes made by its predecessor, the military-historical commission for the description of the Russo-Turkish War, the new military-historical commission was headed by Major-General V. I. Gurko and members who had personally taken part in war, making up thirty-five participants in all.[17]

The majority of the information under the analysis of this new commission consisted of forty thousand pieces *(del')* of documentation generated by the Manchurian army during the war. This enormous collection of information had been arranged at the instigation of the war ministry, which had created a provisional storehouse department of the Manchurian army from the staff of the former active Manchurian army for the purpose of transporting this mass of documents back to St. Petersburg in the spring of 1906.[18] The military-historical commission also drew from other sources, including memoirs, foreign military literature, the reports of foreign military attachés in the Japanese army, and cartographic material from the campaigns.[19] In the period between 1905 and 1908 the military-historical department was also extremely active; its two other commissions were concerned with the evaluation of the Russo-Turkish War of 1877/78 and the military campaign in China in 1900/1.[20]

Also to be subjected to Palitsyn's reappraisal was the reformulation of strategic plans for both the eastern and western frontiers to take into account the altered strategic position of the empire. This process began in October 1906 with the submission of a proposal by Alekseev and Colonel Dobrorolskii which sought to address the composition of the Russian armed forces with regard to the domestic defense of the country in spite of the fact that these armed forces were in no way capable of living up to the international political status of the Russian Empire.[21]

Apart from these various departments and special military-historical commissions set up under the auspices of the chief of the main staff, there was a great expansion in the number of official forums which dealt in some way or another with the analysis of the recent combat experience, the development of theory in light of this experience, and the enactment of appropriate reforms and reorganization in the Russian armed forces (army and navy). One of these was the council of state defense created in May 1905 under the chairmanship of Grand Duke Nikolai Nikolaevich to encompass the general discussion of defense matters relating to both the war and naval ministries.[22] Other centers of military-scientific research which participated in the analysis of the Russo-Japanese War were the staffs of the various major frontier military districts, particularly the Vilna military district which issued a collection of materials on the Russo-Japanese War and then in the following years produced a scientific analysis of the history of the war.[23]

On a more popular level of discussion was the founding of the Official Russian Military-Historical Society (IRVIO) in 1907. Its precursor had been the Society of the Enthusiasts of Military Knowledge, established in 1898.[24] Before the Russo-Japanese War such an official society had been contemplated by the chief of the military-educational archives of the main staff, A. Z. Myshlaevskii, as an alternative to the few existing forums for military-scientific research at that time. However, both the nature and outcome of the war and the need for an organized, officially sanctioned institution to raise the ideological level of the Russian army suggested a program which would "illustrate the most courageous events of our military history and in general to create spirit and strengthen the spiritual rules in the army by means of historical education,"[25] while deemphasizing more specialized and scientific issues of military affairs. It consisted of, beside senior members of the tsarist bureaucracy, many Russian academics of general and military history as well as the military-historical department and the main administration, librarians from other archives, and workers from various museums and libraries.[26]

The society established regional chapters in the large cities of European Russia and decided to divide the range of topics to be addressed by the society into sections, the first being the history of Russian wars, the second being the history of military specialities which consisted of military art and types of troops, and the third consisting of military archeology and archeography for the maintenance of museums and archives.[27] Other than publishing the *Journal of the IRVIO* and popular brochures celebrating the anniversary of significant Russian military events intended for general readership in the armed forces, the society was responsible for the compilation and publication of the *Centennial of the War Ministry, 1802–1902* (*Stoletiie voennogo ministerstva, 1802–1902*) and the *History of the Russian Army and Navy* (*Istoriia russkoi armii i flot*) under the editorship of Skalon.[28]

Despite the enormous increase in the number of official forums for the development of the many aspects of military science, the theoretical work and training provided by the Nicholas Academy remained the central and most significant. With the commencement of discussions over the strengthening of the Russian army's organization, sharp criticism from the press was soon to focus on the argument that obsolete theory and inadequate training at the Nicholas Academy was perhaps the most significant factor contributing to the defeat of the Russian army, illustrated by the observation made by P. A. Geisman that "the failures of our last war aroused attacks from literate society and much noise from the press on the decline of military science, higher military education, and the Academy as well as the whole group of its professors."[29] In response to this sharp criticism, Palitsyn appointed two research commissions to examine the possible faults of the academy's curriculum in conjunction with the developing trends in military-scientific affairs, one within the general staff at large chaired by General Ermolov, and the other, more significant one, in the Nicholas Academy under the direction of its director, General N. P. Mikhnevich.[30] In March 1906 the commission at the academy sent out three hundred questionnaires, receiving only twenty responses by September. Confident that these were representative of attitudes among general staff officers, Mikhnevich established a commission to review the implications suggested by the responses: a referendum on the entire military-scientific process over the preceding twenty years of the Nicholas Academy.[31]

Many of the responses blamed the defeats of the Manchurian army on the incompetent leadership of Kuropatkin or the less sophisticated war matériel of the Russian army, but all the respondents recognized that there were significant changes in the nature of warfare. As one respondent, Colonel V. B. Kol'shmidt, observed in his critique of the Russo-Japanese War, "no previous war gives so much rich material for the study of the actions of large armies in the contemporary significance of that expression."[32] At the most general level comments were made on the need for a general plan of state defense[33] which would provide for more adequate fortification of the theater of war and systematize the assessment of an opposing nation's military capabilities. Given these wider parameters, strategic planning and command of increasingly larger bodies of troops would not only be more coherent and easier to execute, but would also attach a sense of political legitimacy to the war effort and would, in turn, strengthen the resolve of the officers and troops.[34]

Another aspect of modern warfare apparent to these veterans was the increasingly complex role of the general staff in the administration and command of troops in battle. The predominantly clerical (*kantseliarnii*) methods practiced by the general staff in the past which typified Kuropatkin's command in the recent war were no longer sufficient to secure victory in armed combat.[35] As controversial theorist and participant in the war Martynov was to later claim in his work, *The Sorry Experience of the Russo-Japanese War* (*Iz pechal'nogo opyta russko-iaponskoi voiny*), the taste for detail at the expense of the whole, the fascination for generating volumes of paper and designing complicated plans with elaborate diagrams and extensive footnotes immediately invalidated by the rapidity of Japanese offensives, was the very root of the problem.[36] Predilection for the finer points of military administration was a sign of the careerist, as opposed to a talented commander, a product of an overly theoretical education and rapid advancement through the ranks by service in "small" wars.[37]

A few months before his death Dragomirov, in a conversation with the French military attaché, had elaborated on these characteristics and argued that Kuropatkin was unsuitable for command of the Manchurian army for the reason that he was a "Tashkenets":

> A Tashkenets is an officer who, from the point of view of technique has been spoiled by the experience of small expeditions in Central Asia where the adversary is not a formidable warrior, where tactics takes on a secondary priority to the organization of [supply] trains, the sanitary service, and the hygiene of the troops, which are the most difficult and important. . . . Our Turkestan is a vast factory for upstarts.[38]

There was no doubt that Kuropatkin had considered the Japanese less than a serious foe in the months before the outbreak of the war. When he came back from a diplomatic trip to Japan in the summer of 1903, he demonstrated a significant incomprehension of the military threat the Japanese army posed for Russian interests in southern Manchuria. He was reported to have observed: "If we begin a war with them [the Japanese] it will be like war with Turkestan."[39] In the subsequent conduct of operations, his inability to command forces against a sophisticated opponent revealed itself in the practice of dismembering organic tactical formations into ad hoc detachments to counter Japanese responsiveness in the field.

Another veteran and colleague of Martynov to comment on his clerical tendency was Lieutenant-Colonel Neznamov. Recently appointed adjunct-professor at the Nicholas Academy, Neznamov pointed out that the wide use of the field telephone allowed the supreme command to meddle in the more specific command and administrative affairs of his subordinate officers, distracting him from the wider responsibilities of his central command. Focusing on this misdirection of responsibility in the senior command, Neznamov resurrected the Geisman–Martynov–Petrov debate over the growing democratization of command in modern armies with the comment that "the unity of combat will not be possible in the future unless under the unity of a common goal which is accurately and rationally understood by everyone."[40]

The secret to victory in modern war lay in the pursuit of *orientirovaniia,* the hour-by-hour management of information concerning evolving circumstances and responsive instructions—vertically between the central command staffs and subordinate command staffs, and horizontally between the staffs of adjoining regiments and detachments. Duplicating this same concern for the reform of general staff duties, respondents to the commission's questionnaire pointed to a need for "regularizing staff duty and duty of managing the war, the elimination of clerical methods of leadership and the simplification of staff documentation."[41]

Many of the respondents also demonstrated a concern for the low technological sophistication of the war matériel used in the Russian army in comparison to that of the Japanese army, complaining of numerous instances of equipment malfunction which, in the case of radios and field telephones, disrupted communications and forced commanders to revert to older means of communication.[42] More significantly, however, there was awareness of the need to better understand the direct relationship between all types of technological innovations in armaments and their most advantageous tactical employment against the opponent. As Major-General M. V. Alekseev noted in his questionnaire, the main deficiency in the Russian officer corps was that "in many cases officers were not familiar with contemporary tactics of artillery or with quick-firing cannon, guns, howitzers, or mortars."[43] This strongly implied that the tactical forms of new technologies specific to a branch of the service, such as artillery, had not been coordinated with those of other services, nor were they understood by a senior commander trained in the tactics of a different era.

All of these trends in modern warfare pointed to deficiencies in Russian higher military training which had been endemic in various military-educational institutions since the fundamental reform of curriculum in the 1860s and 1870s. Describing the Nicholas Academy as "a breeding ground for military science," these former students criticized the curriculum for having overindulged in military history of the distant past and for having become a classical gymnasium for bureaucrats.[44] More severe criticism of the Leer–Dragomirov theoretical heritage fell to Martynov, who once again demonstrated rare form in denouncing this system which discouraged the development of independent initiative for a more pedantic professional ethic. In a tirade against the excessive scholasticism of the courses on military art, Martynov mocked the reduction of subjective concepts into meaningless mathematical formulas which obscured simple, accurate descriptions of contemporary trends in the theory of military art, and the tendency for excessive classification: "The Academy has even reduced [the concept] of man himself into 'individual-man, social-man, volitional-man, intellectual-man, moral-man, physical-man.' What a salad! In the final analysis does the synthesis of all that create an actual man?"[45] Although long on criticism of a pedantic educational system, Martynov's recommendations paralleled those of Mikhnevich's commission after a review of the opinions stated in the questionnaires. Education of general staff officers was to relate theory more closely to the operational necessities of a modern army with rejuvenated courses on strategy, tactics, and military history. More technical aspects of such issues, such as the mobilization of troops and their transport by rail, the supply of artillery and engineering projects, and the increasing complexity of command, were also to receive increased attention in the curriculum.[46]

In the years following the resolutions of the special commission at the Nicholas Academy, many of its participants continued to evaluate the recent military experiences of the Japanese war in the light of the proposed recommendations for the reform of the Russian armed forces. Much of this evaluation took place in the Nicholas Academy in the form of published essays by individual officers and the collective effort of the academy in a series of conferences held during 1907/8.

Although there was a general consensus that the Russian imperial army was lacking in efficiency and effectiveness, there was little specific agreement as to what was the cause of defeat and what was to be the nature of the subsequent reform. For the most part the contending parties drew up along the well-established lines which had traditionally separated the academic theorists from the nationalistic practitioners, although now the argument was to take a slightly different direction. In the conferences new proponents for a unified theory of military art argued that the most obvious lesson suggested by a review of the war was that there was no common basis for action within the structure of the Russian armed forces, especially in the high command.[47] The conclusion drawn was that rather than proving Leer's teaching to be obsolete the war with Japan had shown how poorly his teachings had been assimilated by senior commanders and how necessary they were for future reform of the imperial army. As Colonel Novitskii, convener of one of the conferences, noted:

The contempt for the science of war in a large part of the army and the possession of this science only by a small number of persons removed from the life of the troops has created conditions very unfavorable for perfecting our military system. One must fight against this situation and spread to the greatest extent possible healthy ideas on the command of troops and those already established by the science of war.[48]

Ultimately, for a majority of the young staff officers who had participated in the war against the Japanese the challenge presented by defeat was to be met with aspirations for perfection suggested by scientific evaluations of the last war.[49]

Perhaps the first independent and most articulate attempt to adjust existing Russian military theory in the light of the experiences of the recent war was by Colonel Gerua who, in his *After the War: About Our Army* (*Poslie voiny: O nashe armii*), published in 1907, suggested that whereas in the past the orientations of Leer and Dragomirov had been separate and exclusive the nature of contemporary war manifested in the Japanese war suggested that these two orientations actually complemented each other.[50] It was his opinion that Leer had done a very good job in establishing the foundation of theory, but had neglected the means by which theory was to be translated into action. Dragomirov's preoccupation with the conduct of warfare within the context of the circumstances of time, space, and the spiritual element of military affairs in his published works and lectures had, in turn, neglected the more abstract considerations of contemporary strategy. Unfortunately for the development of military science in Russia, these two theorists of strategy and tactics, by limiting the scope of their work to separate aspects of military affairs, had ignored rather than developed the natural connection between abstract theory and its application. The development of a truly scientific basis for the conduct of military affairs could not afford to ignore the work of either of these two teachers, for each had contributed significantly to the development of military science.[51]

Drawing from a model provided by a theoretical study written by a professor at the Berlin War Academy, Gerua suggested that the curriculum of the academy be divided into two parts consisting of purely theoretical and applied aspects of military art, and noted that the academy already presented the theoretical part by teaching the criticohistorical method.[52] The structure of theory in the curriculum should follow Blume's program: an introductory discussion on the essence of wars focusing on the means of war and their combined use by the armed forces. Even the much-criticized teaching of the military experiences and genius of the great commanders was justified by Gerua as being the best way to present the teaching of abstract strategic principles. This part of the program was also to be expanded to encompass the experience of more recent European and Russian wars. Such wars provided more purely technical (normative) examples for the potential conduct of warfare in a contemporary theater of war. Furthermore, Russian wars would also provide Russian officers with useful training assuming, as Gerua did, that "military art is general and international, but the merits and insufficiencies are national and therefore inherited."[53]

The applied part of the curriculum was to cover the potential application of these principles to modern circumstances, in the context of mass armies, improved means of mobilization and supply permitted by railroads and conscription, and new means of command. This transition from general principles to application was to be accomplished by the formulation of rules or norms which were the manifestation of eternal principles in these modern contexts and were not in themselves eternal but very much dependent on social, economic, technological, and environmental circumstances.[54]

The operating assumption in the determination of these more normative guidelines for the conduct of contemporary warfare was Gerua's awareness of the inevitable development of a nation's culture accompanied by development of technology. In the case of tactics, this was embodied in the technological improvement of weapons, and for strategy, the development of railroads, mass armies, and command functions. It was in this part of the course that Gerua envisaged most of the work because it was obvious that a course on applied strategy was to be a synthesis of all the auxiliary military sciences and military knowledge.[55]

The goal for such an interdisciplinary research effort was to collate material for critical generalization to be used by strategists and "technologists" to resolve questions in military affairs without resorting to improvisation. Gerua recognized the potential problems in the provision of this more recent and refined research process as well as the danger of losing the focus of main trends in the evaluation of military art due to the mass of detailed information, but emphasized the existence of an already partially collated field of information for application.[56]

Much to Gerua's dismay, an essential component of the applied section of the curriculum—the conduct of field exercises in tactical and strategic maneuvers, war games, and the development of strategic operations—was being neglected in the training of officers.[57] In biting criticism of the ad hoc and unsystematic practice of strategic maneuvers on the part of military district staffs, Gerua pointed out the effect, or lack of effect, such neglect of field experience had on the performance of general staff officers in the battle in Manchuria, removed as they were from the actual maintenance of the army or needs of the troops.[58] He emphasized his point with a caustic description of a strategic exercise in which he had participated to emphasize the need to improve the actual conduct of the exercises and to perform them systematically and biannually.[59] In a concluding remark on the growing role of the general staff in modern warfare, Gerua once again emphasized the need to build a bridge between theory and the nature of modern combat for success in the field of battle, an endeavor which would greatly benefit from the excellence of both traditions in the Russian army. Throughout the remainder of the interwar debate over the reform of the curriculum at the academy, various professors and military theoreticians were to devote much effort to the development of this synthesis between theory and practice upon the well-established military theoretical legacy of Leer and Dragomirov.

In 1911 Mikhnevich, former director of the academy and acting commander of the 5th Army corps, issued a third edition of his famous text on strategy after significant revision in light of the experience gained in the war with Japan. As with all texts concerning the study of strategy since the publication of Leer's classic, *Attempt at a Critico-Historical Method* in 1868 Mikhnevich's work reviewed the development of general knowledge and what the implications of this progress were for the development of a specific military science.

The father of scientific positivism, August Le Comte, had asserted that man's knowledge had evolved from the inspiration of individual genius through various stages of sophistication to eventually approach the ideal of positive knowledge (an applied science) practicable by all. With this evolution of knowledge existed two fundamental paradoxes. On the one hand, there was the well recognized contradiction between the purely theoretical and purely applied aspects of science and on the other, the static and dynamic nature of empirical reality.[60]

Mikhnevich, like Gerua and Clausewitz before him, portrayed the first paradox as the rivalry between men of ideas and men of action. Intending to appease the strong nationalist sentiments within the military community which valued the subjective, spiritual aspect in the evolution of military art, he praised the role of great military geniuses, attributing to them new insights into the nature of warfare and enactment of innovations in strategic and tactical organization. However, without continual evaluation of the methods to guide the resolution of problems in military affairs and familiarity with the causes behind the evolving manifestation of eternal principles in strategy and tactics provided by the pursuit of military theory, the man of action was highly prone to methodism.[61] Victory in modern warfare was now determined by the degree of symbiosis between the two, the ideal military science, and neither aspect could be ignored: "In any case the discord between rational theory and experience does not have to exist; between them there must exist strong ties and for this reason there cannot be any question of which is more important."[62]

In his discussion of the second paradox which existed between the eternal, unconditional principles regulating the dynamics of social phenomena and the empirical manifestation of these principles, Mikhnevich came to the defense of the greatly abused criticohistorical method in studying the changing physiognomy of military art. The academy's curriculum, founded on the evaluation of military statistics and the study of the evolutionary development of eternal military principles over time, coincided with Le Comte's distinction between static and dynamic components in the study of sociology.

Despite the many disagreements among subsequent social philosophers over the principles which determined the nature of evolution in social phenomena, all of them had shared the fundamental opinion that "social evolution perfects itself in a regular manner, i.e., that separate sides of social modes of life develop in defined order," and that war as a social phenomenon would also follow this evolutionary pattern.[63] Consequently, there was no question of the evolution of military art taking a turn for the worse as was felt by many European and Russian military theorists before the turn of the century. This relationship between the cultural development of the nation and its evolving military art was the first to become apparent in the conduct of tactics which would, in turn, influence the conduct of strategy and, ultimately, politics. Nor was this interconnection between civilization and military art to be the exclusive property of military theoreticians; comprehension of this evolutionary process in military art was deemed by Mikhnevich as essential for the military commander's proper understanding and manipulation of various circumstances existing on the battlefield: "without a hand on the 'historical pulse' defining the elements of a nation's forces at a given time the commander will not master sufficient wisdom in defining those circumstances which exist in the given war."[64]

In coming to terms with the nature of modern warfare Mikhnevich felt, as did many others on the general staff, that the combination of conscript armies with rapidly advancing technologies to transport them quickly over greater distances and to furnish them with greater firearms would completely exhaust a nation's material and spiritual forces. More than ever before the conduct of warfare had become an exercise in "economy of force" which required full coordination of governmental policy and dislocation of troops on the field of battle as well as the diplomatic effort in maximizing the strategy of a wartime coalition. Given the increased potential to dedicate the nation's resources to a confrontation involving many national armies over an extended duration of time, this economy of force could be best provided by the decisive leadership of a monarch. Governments where executive power was diffused in various political institutions or which found themselves in possession of inferior military forces or lack of the strategic depth necessary to conduct an extended defense had to rely on quick success in the form of a general engagement at the beginning of hostilities.[65]

Borrowing from the work of Bliokh and Gulevich in the determination of a nation's cultural integrity, Mikhnevich concluded that the Russian Empire at the beginning of the twentieth century was still in a "natural" stage of cultural development in contrast with the more sophisticated "timocratic" cultures on its western borders and the "semiprimitive" cultures of Central Asia. As a centrally located, predominantly agrarian culture, the Russian Empire had been obligated to develop diplomatic and military policies suitable for its relations with both worlds. In the international arena the implications of this were clear: both Bliokh and Gulevich had claimed that in a highly advanced world economy easily disrupted by the stress of coalitional, attritional warfare, Russia's less sophisticated but resilient agrarian economy would provide a stabilizing influence in world affairs.[66] Blessed with an agrarian economy, vast expanses of territory, and a monarchy to lend decisive, cohesive guidance to foreign affairs, the Russian Empire could afford to maintain the military policy of the strategic defensive on its western frontiers as Miliutin and Obruchev had done since the reforms of the 1860s. As Mikhnevich was to conclude, "in any case time is the best ally of our armed forces, and for that reason we should not be afraid to use 'strategic fatigue and exhaustion' to avoid the initial decisive clash with the enemy over our border when the superiority of strength will be on his side."[67]

A final advantage of being an agrarian society was the ability to mobilize men who were spiritually and physically more sound than urban workers because, as

Mikhnevich put it, "town life and work in factories lowers the physical quality of the population making them less able to undertake the burdens and deprivations of [army] life."[68] Later he developed this opinion further within a discussion on the determinants of spiritual qualities of the troops with the explicit statement that "with the development of culture the martial spirit of the nation weakens," but that the young Russian nation would still be able to withstand 30 to 40 percent losses as it had done in the past.[69]

One major disadvantage of such a culture was the comparatively low rate of growth in peacetime rearmament programs in relation to neighboring armed forces.[70] Taking this into consideration Mikhnevich recommended doubling defense expenditures in case the Russian Empire were faced with the disadvantageous position of fighting a large European war without allies. In conjunction with this he also recommended the development of economic plans to supply all necessary populations of the country in accordance with foreign and military policies.[71]

In conjunction with Mikhnevich's wide-ranging and highly abstract synthesis of military theory, a colleague at the Nicholas Academy, Colonel Neznamov, focused his attention on the translation of theory concerning the conduct of modern war into specific innovations in the organizational responsibilities of command in the field. Having graduated from the Nicholas Academy in 1900 and having participated in the Russo-Japanese War, Neznamov was well versed in the legacy of Russian military science and tempered by personal experience with modern warfare. In the introduction to his *Contemporary War: The Conduct of Field Armies* (*Sovremennaia voina. Deistviia polevoi armii*), published in 1912, he acknowledged this intellectual debt by quoting Dragomirov's famous maxim: "Victory goes to that person who best creates a plan of action, takes his army to the decisive point of combat and succeeds in concentrating his forces in the field."[72]

In Neznamov's opinion it was this perennially elusive economy of force on the battlefield which posed an even greater challenge to the supreme commander responsible for the deployment of enormous armies over an increasingly complex battlefield. Whereas in the past the skill of a single talented commander could have led his troops to victory, in the age of the railroad, telephone and radio communications, and rapid-firing artillery and small arms this was no longer the case.[73] Amidst technological and quantitative change the role of the subordinate commander had gained significance, as was apparent in the military operations during the Manchurian campaign which had been seriously disrupted by the incompetence of unit commanders.

This recent experience demonstrated the fundamental paradox of conducting warfare in the modern era: the need for a single direction in the conduct of operations was stronger than ever but the individual commander who once provided that direction on the field of battle had been removed from direct command. Regarding the army as a single mechanism where the function of one part must exactly conform to the activity of the other parts, Neznamov insisted that there be a unified military doctrine which would facilitate explicit communication in the exchange of orders and implicit communication through standardized military training:

> The more sophisticated and unified the preparation of all command personnel, the more fully will the idea of the commander be executed; the more he can be sure in the correct understanding of his idea by his subordinates and the more they are skilled in their specialities, the more audacious can be the plans he makes for himself. For that one needs full communication.[74]

Playing a very important role in the formulation and continual reevaluation of this doctrine was military theory which, by exploring the conduct of recent war and delving into the vast accumulation of ideas throughout military history and from every civilization, would evaluate contemporary forms of warfare and even permit a glimpse of the possibilities in the near future.[75]

This new orientation to theory and its direct implications for the professional ethic of the Russian army drew criticism from all sides, not least the advocates of the national tradition in Russian military art who perceived a foreign-derived theory which threatened traditional leadership values personified by Suvorov and Skobelev. In response Neznamov acknowledged the value of this tradition by commenting that "we, by nature of [our] Slavic nature are very inclined to lay all our hopes on the natural ability of talent and deeply believe that 'war gives birth to heroes.' " Yet innate talents were no longer sufficient: one also had to be fully versed in the modern science of warfare.[76]

Distressed by the poor strategic guidance of the war with Japan and the humility of defeat, officers on the general staff once again turned their attention to the formulation of war plans for the coordination of the empire's defense. In October 1906 the head of the operations department of the newly established main administration of the general staff, Major-General M. V. Alekseev and his colleague Colonel Dobrovolskii drafted a proposal to the chief of the general staff, calling for the compilation of a general plan of state defense. This new component was to assess the threat to the weakened Russian Empire posed by various countries or coalitions, such as the Triple Alliance on its western border and the coalition of Japan and Great Britain in Asia, and then determine the most effective composition of the Russian army and fleet while they underwent a restructuring of their forces.[77] Recognizing the increasing dependence of military policy on other ministries within the government, the proposal listed the following priorities:

> The following measures were proposed for the development of the armed forces: to abolish reserve forces and maintain during peace time an army existing exclusively of combat troops; to technically improve the army's equipment with new weapons (machine guns, etc.); increase the size of the armed forces during peace time; introduce a territorial scheme for building up the army; make broad preparations for the theater of war using

engineering methods; devise a plan for developing a railroad network in border regions; more intensively use the systems of highways, dirt roads, and waterways; create an active navy and improve the well-being of the masses by raising their level of education which should improve the make-up of the armed forces.[78]

From its inception in 1906 to the proclamation of the Great Military Program in 1912/13, this proposal was to provide direction to the reform of the Russian army and set the main administration's agenda for strategic planning.

In these years of reform and institutional debate in the general staff, the issues involved in strategic planning became the subject of intense theoretical development by professors at the Nicholas Academy, among them N. P. Mikhnevich, A. A. Neznamov, V. F. Novitskii, I. Maslov, and B. Koliubakhin. Commonly accepting the notion that the pursuit of perceived national interests by military means could quickly mobilize and transport men and material to the battlefield, these theorists recognized in the preparation and unfolding of war the ultimate expression of the eternal principle of "economy of force." One side of this economy was the ideal of warfare or strategic offensive at the outset of a war: "Military science using the many centuries of historical experience teaches us that for victory one needs to be stronger than the opponent at the decisive point and at the decisive moment."[79] Assuming that the political goals pursued by the belligerents did not outweigh the military losses, the army which sought and obtained victory in the first decisive battle had entered upon a progression of military operations which were relatively simple compared to those defensive actions thrust upon the opponent: "conditions of space and time are only means to annihilate or weaken the enemy," whereas defensive battle was prone to the vicissitudes of circumstance.[80] Within a few years after the conclusion of the Russo-Japanese War, defeat upon the commencement of a campaign had become an implicit concern within the Russian imperial general staff. During a visit of General Sukhomlinov to a lecture by Neznamov on the liabilities of the *operation de couverture,* or screening operations used by the Russians against the advancing Japanese army along the Ya-lu River, the war minister was reported to have commented: "Yes, no more Tur-en-chens, there is no need to repeat that mistake."[81]

Although this ideal was seen to be an eternal principle of warfare, its manifestation in the era of the "nation-in-arms" and the extension of hostilities along enormous fronts demanded that it no longer be expressed as a Napoleonic absolute superiority in a single decisive battle but a local responsiveness to detected threat posed by the opponent over a succession of battles.[82] Ultimately the task of the senior commander was to differentiate these theaters of war, or sectors of battle, where the forces under his command would assume the offensive, primary goal of gaining decisive victory in battle from those less important theaters of war where he would assume defensive, secondary goals by trading time for space with maneuvers designed to disrupt the unity of the opponent's offensive.[83] In practicing this delicate calculus, or *glazomer,* for the purpose of pursuing the offensive, the commander had recourse to analyzing his current circumstances according to several eternal principles in the form of questions concerning the correlation of superiority, security of supply, unity of communication, and responsiveness of command to achieve surprise over opponents:

> 1.) How to form sufficient preponderance of force in the main Theater of war? 2.) When a preponderance has been reached, how does one destroy the force of the enemy who preserves his [strategic] preponderance to the end of the struggle? 3.) How to construct unbreakable communications between the field army and its rear supply? 4.) How to secure instantaneous and consistent supply of all the army's needs?[84]

The other aspect of the strategic economy of force was the recognition of inherent limitations of a nation's actual military capabilities which might determine a relatively inert deployment of strategic forces. In the case of the Russian army, the geographical expanse of the borders it was to protect in conjunction with its comparatively underdeveloped system of railroads dictated the temporary abandonment of the immediate offensive in favor of the strategic defensive:

> The weaker side in the given case does not refuse the offensive, he waits for it, but not being prepared for it waits and takes measures which will slow and limit the success of the opponent. He goes on the strategic defensive which can be seen as a continuation of preparations not completed in peace-time for decisive battle.[85]

A very important component of the strategic plan which had developed out of the general plan of state defense was the semistatic configuration of supply lines and depots provided by the integration of railroads and fortresses. Although railroads had proved to be the most significant strategic technology available to modern societies and played a significant role in both the preparation and conduct of battle, their track, tunnels, and bridges were particularly susceptible to physical destruction by the enemy. Their linear, sequential limitations of transport itineraries could be easily disrupted by unanticipated changes in circumstance as had been demonstrated by warfare with the Turks and the Japanese.[86]

Given the fragility of these powerful arteries of supply so important to contemporary war, the strategic significance of fortresses underwent a dramatic change. No longer were fortresses limited to the function of a strong supply depot for the field army and built with heavy fortifications and defended by heavy fortress artillery. They were now to fortify the integrity of the railway's flow.[87] This geographical system of strategic technologies and defenses lent a quality of rigidity to the choice of strategic goals and the subsequent deployment of forces to the field, making strategic planning less the inspirational expression of the senior commander and more a scientific, systematic expression of the entire state's military potential.[88] During the First World War this hypothesis would be substantiated by a Russian military theorist and observer on the Russo-German front who remarked that "Our present war gives an unparalleled model of the use of railroads, but at the same time this single artery which connects us with the base cannot but influence the essence of our strategy, which is dominated by the principle: maintain that artery."[89]

One of the more significant lessons to be drawn from the war with Japan was that preparations for the initial general engagement between technologically advanced armies were no longer sufficient to gain victory. The historical circumstances which had given birth to the Napoleonic ideal of an uninterrupted offensive followed by a rout of the enemy forces had succumbed to the era of the nation-in-arms, where armies were continuously reinforced throughout the duration of hostilities. This new attritional nature of warfare which consumed highly trained manpower and sophisticated weaponry at tremendous rates was punctuated by the increasingly frequent phenomenon of periods of calm in the midst of battle during which supply and regrouping of forces for a renewed onslaught were conducted.[90] In light of this cycle of action and preparation for action, new responsibilities were required of the commander and his staff: the conduct of a series of successive military operations against the opponent's armed forces in pursuit of the larger, strategic goals of the war.

Reevaluating the existing theory of military art to account for this phenomenon of contemporary warfare, several military theorists at the academy turned their attention to the relationship between the senior and subordinate command in the field army. As had been evident in the Russo-Japanese War, the major problem in the command of multimillion-strong armies was the threat of administrative disorder in the delegation of authority and transmission of information between different levels of command, compromising the responsiveness of the subordinate commander to local circumstances.[91] This was equated with the process of democratization in the practice of military art which had been brought about by technological advances in communications. In order to approach the analysis of this problematic relationship these theorists adopted the theoretical-balance legacy which Dragomirov and Leer had developed in the 1860s and upon which they had constructed their theory of military art. This new level of theory in military art would seek to systematize the balance of the subjective role of the subordinate commander with the integrating role of the senior commander.[92] The central concept of this new level of military art was *edinonachale,* which was to assure that the conceptual goal set by the supreme command at the outset of hostilities was maintained throughout the successive military operations. In part this conceptual unity of the Russian army was to be provided by the designated commander-in-chief of the field armies and his supervisory staff, or *stavka,* modeled on the supreme staff of the German army.[93]

On the other hand, the subordinate commander at the front was to be given more freedom in planning and executing responses to circumstances specific to his locale. This consisted of identifying the goals of the operation in concert with the strategic goals of the war, determining the first task and method to be used in reaching that goal, and then taking measures to maintain the rate of reinforcement required to achieve those goals.[94] More circumspect than strategic planning because of its inherent proximity to the evolving circumstance of the battlefield, the conduct of operation planning placed a high premium on the collection and analysis of information gained from many intelligence-gathering sources. In peacetime the collection of predominantly indirect information relating to the military power of neighboring nations and member nations of coalitions was performed by the main administration of the general staff and the frontier military district staffs, and consisted of the mapping out of the theater of war, statistical and historical information, communiqués from Russian military attachés at embassies in foreign nations, and espionage. In wartime this was supplemented by direct gathering of information by staff organs throughout the hierarchy of the field administration of the army, and consisted of tactical information gained through long- and short-range cavalry reconnaissance, encounters with the enemy by the infantry and combined-arms detachments, and aerial observation from balloons.[95] Based on this constant flow of information and analytic procedures, the subordinate commander and his staff could preserve freedom of movement—regardless of the material correlation of forces with the opponent: "The more in detail the goals are described, the more preparations can be made, the less chance will influence the future and the more energetic and decisive one can act."[96]

The personal role of the subordinate and senior commanders in maintaining this unity of command was through the traditional practice of *glazomer,* which had evolved from being merely an instantaneous evaluation of battlefield circumstances to *predvidenie,* or foresight into the possible immediate actions of the opponent.[97] In the opinion of Neznamov, the commander should not place himself in the shoes of his opponent by attempting to foresee his actions but only conduct a worse-case scenario of his own position as had been taught in tactical map exercises and maneuvers since the 1880s and 1890s: "A good commander should ask himself throughout a day 'what do I do if the enemy appears on my front, my left or right flank?' "[98]

This was the only reliable approach to the commander because the ephemeral nature of reconnaissance concerning the activities of the opponent would never be complete enough to successfully guess the opponent's intentions, nor did one need to know if one remained aware of all the possibilities. Because the nature of information was of marginal, short-term value and an army's operational responsiveness was inherently compromised by administrative inertia, Neznamov emphasized both the need for more reconnaissance of enemy dispositions and more efficient internal lines of communication.[99] Accuracy of information and responsiveness of organization could also have a positive effect on the morale of the troops so fundamental to the conduct of contemporary warfare: "certainty gives the stubbornness to hold on."[100] An important aspect of this close interaction between information and the responsiveness of field administration which advocated continuous and accurate analysis of reconnaissance was the increasingly powerful ploy of disinformation and secrecy:

The more unexpected our actions are for the opponent the more difficult it will be for him to take measures to parry those unexpected things he did not foresee; the more difficult it is for the opponent to collect information about us the less perfected his preparations for operations will be; the later he knows about one or another of our measures, the less time he has for correcting his mistakes, the less order he will have in his actions, and the more hope that he will not be in the position to commit all his forces to the task.[101]

The other and equally significant aspect of *edinonachale* was the recognition on the part of the senior commander for the need to delegate increasingly greater command responsibility to subordinates in the interest of maintaining local initiative and to entrust greater technical responsibilities in the everyday administration of the army to his staff, freeing him to attend to his unifying role as senior commander. Unlike the opinions of military theorists before the Russo-Japanese War, this was no longer considered a manifestation of dwindling creativity in the role of the commander, or a fall in the level of military art:

In our time one hears more about the lessening influence of the supreme command on the outcome of battle. "Today's battles are not the battles of generals but the battles of captains"–state a few French writers. Of course, this is not true. General Lewal's opinion is hardly sympathetic with this: this opinion is very much the opinion of the decadent epoch, the "fin de siecle" mentality.[102]

To counteract the growing complexity of command, the general staff was to cast off its traditional passive clerical duties and adopt a more active role in assuming the technical responsibilities of command, such as the collection and analysis of information during wartime and the composition of orders and directives of the commander and other specialized technical services involved in the day-to-day administration of the field army, allowing the commander to fully focus on his creative responsibilities.[103]

The other aspect of *edinonachale* in conducting military operations was *orientirovaniia,* the concept first formally recognized in Russian military theory in Geisman's work in 1893 and reiterated in Neznamov's evaluation of the Russo-Japanese War.[104] On the tactical level the maintenance of communication for coordination of combined arms operations in the vanguard of the army was relatively easy: "The presence of the division commander at the front, reacting to the evolving circumstances of the battle allows him personal orientation with other divisional commander due to geographical convenience."[105] The problem of maintaining "lively interaction" between different commands in the army existed between the commands of the higher troop formations where personal interaction would be impossible. The only way to maintain this lively interaction would be via the general staff representing the will of the senior commander at staffs of subordinate military formations.[106] In the opinions of these military theorists who strove to understand the nature of contemporary warfare it was this new professional ethic which shunned the clerical tradition of staff work to advocate a more active, technical role in the administration of command which represented the true revolution in the development of warfare.[107] What was needed to maintain this harmony of command both in the explicit and implicit administration of command was a "single military school, one method and manual for resolving military questions"—a truly unified military doctrine.[108]

Ultimately the purpose of all those theories and preparation was to emphasize the role of tactics as the basic element of all military affairs.[109] In taking into consideration the problems of space for maneuver in the contemporary theater of war, Neznamov felt that the initial clash would be purely frontal, reminiscent of Petrov's prediction that modern wars would resemble brutal battles between rams.[110] The conduct of battle was seen as a microcosm of the war in general and within the realm of this they implied that the nature of modern battle also could be defined as offensive, where "the conditions of space and time are only means to annihilate or weaken the enemy," or as defensive, where circumstance would determine how secondary goals were to be sought in the disruption of the opponent's unity of action.[111] Naturally the ideal was the offensive because it was not as acutely determined by circumstance and thus more easily allowed a unity of purpose so essential in the conduct of large armies to permeate the strategic plan and subsequent operational plans.[112]

Concerning the actual mechanics of combat tactics, Neznamov owed a great debt to the theoretical legacy of Dragomirov despite the fact that he advocated fire-power over cold steel and intelligence over will. The mechanics of modern infantry tactics in conjunction with artillery and rifle fire consisted of a theory of balance between the spiritual perseverance of the soldier throughout the carnage to give the final blow and the efficient coordination of combat formations to exercise effective economy of force in gaining numerical superiority at the most advantageous time and place.[113] During the phases of the approach, the attack, and the final assault during which the senior commanders evaluated reconnaissance, elaborated plans, and issued orders, the troops in combination with artillery to suppress enemy fire would attack in skirmish formation by leaps and bounds.[114] Despite this emphasis on open formation assaults and the predominant role of artillery, rifle, and machine-gun fire, Neznamov still paid tribute to the traditional symbol of shock-tactics: "The bayonet in the hands of a highly motivated soldier has not out-lived its time, only the occasion when it is used is more rare than previously."[115]

By 1908 the slow pace of the Russian army's rehabilitation from the trauma of the Russo-Japanese War induced the tsar to revert the division of authority over military policy back to centralized authority under the responsibility of the war ministry. Nicholas II was to initiate this consolidation by dismissing Grand Duke Nikolai Nikolaevich from his position as chairman of the council for state defense in 1908 and appointing War Minister Rediger as ex-officio chairman, then completely abolishing the position in May 1909.[116] This was followed by the dismissal of General Palitsyn as the chief of the general staff independent of the war ministry and the temporary appointment of Sukhomlinov as chief of a general staff now reintegrated into direct subordination under the war ministry and united once again with the functions of the main staff. By March 1909 Rediger was asked to resign his position as war minister and was replaced by Sukhomlinov.

Sukhomlinov shared Rediger's commitment to building a more efficient, economic, and powerful army; but whereas Rediger was hesitant to bring about major changes which would threaten to disturb the orderly process of reform, Sukhomlinov was more bold. By 1910 he was to embark on a coordinated series of rearmaments, reorganizations, and reforms which were intended to more quickly set the Russian army back on its feet. Sukhomlinov's request for a supplemental appropriation of state funds to finance this coordinated revitalization of the Russian army was to experience very little opposition from either the council of ministers or the Duma.[117] In the internal reorganization of the war ministry which was to consolidate the minister's control over the various collegiate bodies, administrations, and inspectors-general, the main staff and its mobilization section were to be reintegrated into the functions of the general staff.

In fulfiling his coordinated reforms Sukhomlinov, in conjunction with Danilov, secretly worked toward the compilation of a revised mobilization plan (Plan No. 19)[118] which, based on Sukhomlinov's poor opinion of the utility of the immense fortress network on the western frontier of the empire, was to concentrate on the withdrawal of the Russian army to behind the Bug, Narev, Bobr, and Nieman rivers where full concentration of the Russian army could take place for the commencement of offensive actions against East Prussia while maintaining a defensive posture against the Austrian front.[119] Although finance was forthcoming from the council of ministers and the Duma, most of it was allocated for the renovation of fortresses and construction of highways—projects which detracted from Sukhomlinov's concern for the development of the field forces. Seeking an alternative program which would meet these demands, Sukhomlinov appointed General M. A. Bel'aev to head a commission in the general staff which was to eventually compile the *Bol'shaia Programma,* presented to the war minister in the summer of 1913.[120]

However, this material augmentation and organizational reform would lose much of its significance in the absence of a corresponding concern for the increased sophistication of the officer corps' ability to conduct modern warfare. To counter this problem Sukhomlinov promoted a reform of the military-educational system in which the Imperial Nicholas Military Academy (the new name of the Nicholas Academy of the General Staff after 1909) under the directorship of General Shcherbatchev was to play a central part.[121]

Unimpressed by the academy's overemphasis on abstract theory and military history to the detriment of training general staff officers in the fundamental, practical aspects of their duties, Sukhomlinov sought to transform the curriculum to the French training system used at the *École Supérieure de Guerre.* To accomplish this he requested the aid of the French general staff which, in turn, dispatched Colonel Janin to supervise the adaptation of French methods in October

1910.[122] Sukhomlinov's aspirations were not only to improve the practical training of general staff officers but to make the Imperial Nicholas Military Academy the intellectual center of the Russian army by incorporating its teachings on military affairs into the new field regulations being formulated at the time, thereby guaranteeing a unified training to the entire Russian army.[123]

In a commission chaired by General Ruzskii under the department of regulations and duty of troops within the main administration of the general staff, professors from the academy were to help compile the field regulations to be published in April 1912. However, during this time there arose a disagreement among the compilers as to whether the commission should formulate a unified military doctrine in these regulations.

The first to champion the cause of a unified doctrine was V. Domanevskii who, in a series of articles in *Russkyi invalide* entitled "On the New Field Service Regulations," "Unity of Military Doctrine," and "Unity of Military Doctrine and the Field Regulations" was to argue that the nature of contemporary war was to impose upon the leadership of an army the reality that direct command of battle would be performed by young, relatively inexperienced officers who would be following only the most general of orders from above. This independence of the subordinate commander was to require that he not only select the proper means to accomplish a set goal, but also "be in the position to act not only without orders, but even in spite of orders which do not meet the situation."[124] In the absence of a command structure that could efficiently and effectively provide direct command to tactical operations from on high, Domanevskii argued that there should be an explicit doctrine in the form of field regulations as well as an implicit one residing in the officer's professional capabilities (trained and innate) which would provide basic guidelines to the conduct of battle. The task, he acknowledged was a difficult one:

> The compilers of the regulations must give to numerous questions separate vital codes, basic fundamental views to provide a point of contact, a general reference for the estimation of circumstances; to formulate general, similar concepts, a *tactical language* which would finally give confidence in identifying concepts, to enclose into one word many meanings.[125]

In an additional comment which closely resembled Mikhnevich's concern for historical and cultural determinants in a nation's military art, Domanevskii fully developed his argument for this tactical language, or linguistics of military art, called a unified military doctrine.[126]

That such doctrine existed and that it was the duty of professors at the Imperial Nicholas Military Academy to formulate and disseminate it via the field regulations to the Russian army was indisputable in Domanevskii's opinion. The dilemma posed by contemporary war required it, and the academy was the only institution in the Russian army which had the time and the intellectual resources to do it: "Life itself pushes us toward taking a decision to give the army a new set of rules for field service. Not only is it lawful, but completely necessary to build into this regulation a military doctrine which has been created through long, selfless, profoundly thoughtful and successive work of the best [intellectual] forces of the Academy."[127]

However, opponents of this unified doctrine soon asserted their position on the incorporation of such a theoretical discipline in the affairs of the armed forces. Immediately after Domanevskii's first articles appeared in November 1911, A. M. Zaionchkovskii, at the time a professor at the academy, wrote an article entitled, "Unified Military Doctrine (Dangerous and Distracting)," denying in no uncertain terms the need for such a doctrine with the assertion that "in the end this is all the same as an attempt to link known theoretical conventions with living art . . . to conform its role into dead and rational formalism which can be very good for theory but always fatal in practice."[128] Zaionchkovskii's views were subsequently supported by articles from M. D. Bonch-Bruevich, D. Zuev, K. Adzaridi, E. Krivtsov, F. Ogorodnikov, A. Borets, and others, who argued that what was needed by the army was a single will, goals, and aspirations, not more scientific formulations.[129]

In reaction to this criticism Domanevskii's argument was in turn supported and augmented by his mentor A. Neznamov, V. Grigor'ev, and others who felt that a distinct doctrine would not exclude but enhance free creativity which, operating by common criteria in their approach to problems, would allow the commander to concentrate his creativity on the details of the battle. Military art for art's sake would lead to anarchy of theories.[130] This polemic continued in the military press well into the summer of 1912, when it was concluded rather abruptly by Nicholas II himself. Perhaps mistaking the subject of the controversy to be an infringement of his prerogative to formulate doctrine in the sense of coordinated civil and military policy between ministries for state defense, Nicholas II informed the director of the military academy that "military doctrine consists in the execution of my orders. I ask you to communicate to Neznamov that he is no longer to publish articles on this question."[131]

A further blow to the aspirations of the professors advocating a unifying reform of the curriculum at the academy came in the spring of 1912, when Shcherbatchev was replaced as director by the pedantic Ianushkevich.[132] In the intervening years before the outbreak of the First World War, Ianushkevich attempted to subvert the influence of the previous reformers and the new techniques borrowed from the French higher military-educational system by reimposing antiquated Applied School methods of tactical instruction.[133] During these two years the administration and curriculum suffered at the hands of opposing faculty members who were struggling to gain full control over the content of the courses. Although efforts to correct this situation were made in March 1914, when General Engalichev replaced Ianushkevich as director of the academy, it continued to suffer a very poor reputation at court and throughout the Russian army up to the beginning of the war a few months later.[134]

NOTES

1. W. D. Bird, *Lectures on the Strategy of the Russo-Japanese War* (London, 1909), 29.

2. G. Donnet, *Histoire de la Guerre Russo-Japonaise* (Paris: Delagrave, 1904/5), 13.

3. *The Russo-Japanese War,* Prepared in the historical section of the German general staff, translated by Karl von Donat (London: Hugh Rees, 1908) 1: 144–45.

4. Donnet, *Histoire de la Guerre Russo-Japonaise,* 332.

5. B. W. Menning, "Bayonets before Bullets: The Imperial Russian Army, 1861-1914" (unpublished manuscript, 1989), 323.

6. Ibid., 336.

7. Carton 7N 1477, Moulin, "Situation générale de la guerre" (Feb. 10, 1905); "Sur l'arrivée à Petersbourg du Général Dragomiroff" (Mar. 10, 1905).

8. Donnet, *Histoire de la Guerre Russo-Japonaise,* 142.

9. Menning, "Bayonets before Bullets," 341.

10. Although Obruchev was to resubmit his proposal for the reorganization of the main staff to provide for the increasing operational-strategic duties of the modern battlefield and to increase the status of the chief of the main staff along Prussian lines, all serious consideration of reorganization was left to Sakharov, appointed as chief of the main staff in 1900. In February 1900 he was to express the opinion at the military council that "the huge development of our neighbor's armed forces and all their preparations for war demand from Russia extreme exertion. . . . All these demands are directly reflected in the work and structure of the Main Staff" (cited in P. A. Zaionchkovskii, *Samoderzhavie i russkaia armiia na rubezhe XIX-XX stoletii, 1881–1903* [Moscow: Izdatel'stvo 'Mysl', 1973], 110). See also J. Erickson, "The General Staff: Theory and Practice from the Tsarist to the Soviet Regime," in *National Security Policy: The Decision-Making Process,* edited by A. L. Pfaltsgraff, Jr., and U. Ra'anan (Medford, Mass.: Archon, 1984), 20–21.

11. A. G. Kavtaradze, "Iz istorii russkogo general'nogo shtaba," *Voenno-istoricheskyi zhurnal* 7 (1972): 87. See also L. G. Beskrovnyi, *Armiia i flot Rossii v nachale XX v: Ocherk: Voenno-ekonomicheskogo potentsiala* (Moscow: Izdatel'stvo 'Nauka', 1986), 37.

12. Ibid. See also Beskrovnyi, *Armiia i flot Rossii v nachale XX v,* 49: Zaionchkovskii, *Samoderzhavie i russkaia armiia na rubezhe XIX-XX stoletii, 1881–1903,* 110–12.

13. Kavtaradze, "Iz istorii russkogo general'nogo shtaba," 88.

14. Ibid.

15. Ibid.

16. A. Avdeev, "Voenno-istoricheskie issledovaniia v russkoi armii," *Voenno-istoricheskyi zhurnal* 3 (1986): 83.

17. P. A. Zhilin, pod red., *Russkaia voennaia mysl' konets XIX–nachalo XX v* (Moscow: Izdatel'stvo 'Nauk', 1982), 41. When asked whether his commission would compile the history of the Russo-Japanese War any faster than the previous commission, Gurko confidently replied that it would be done in three years, an indication of the importance assigned to evaluating the lessons from this war. Cited in "Sluzhba v Glavnom shtabie i Glavnom upravlenii general'nogo shtaba (1903–1908 gg.)," E. A. Nikol'skii Collection, Hoover Institution on War, Revolution, and Peace.

18. Avdeev, "Voenno-istoricheskie issledovaniia v russkoi armii," 82.

19. Zhilin, *Russkaia voennaia mysl' konets XIX–nachalo XX v,* 41.

20. Ibid.

21. I. I. Rostunov, *Russkaia front pervoi mirovoi voiny* (Moscow: Izdatel'stvo 'Nauka', 1976), 88.

22. Erickson, "General Staff," 22. See also Kavtaradze, "Iz istorii russkogo general'nogo shtaba," 52; Carton 7N 1477, Moulin, "Au sujet du rôle des membres de la famille impériale dans l'armée" (Nov. 20, 1906). Commenting on the work of the council of state defense, the French military attaché recalled Dragomirov's personal disregard for the martial skills of the elder and younger Grand Dukes Nikolai Nikolaevich, accusing the elder of rampant fraud and favoritism during the Russo-Turkish War and remarking that although Kuropatkin had led the Manchurian army in the Russo-Japanese War poorly, he was at least honest.

23. Zhilin, *Russkaia voennaia mysl' konets XIX–nachalo XX v,* 44–45.

24. Ibid., 58.

25. V. A. D'iakov, "O vozniknovenii, sostave i deiatel'nosti russkogo voenno-istoricheskogo obshchestva (1907–1917)," *Problemy istorii doshchestvennogo dvizhenie i sitoriografii* (Moscow, 1971), 276–79.

26. Ibid., 281.

27. Ibid., 284.

28. Ibid. See also Zhilin, *Russkaia voennaia mysl' konets XIX–nachalo XX v,* 63.

29. P. A. Geisman, "Dopolnenie k stat'e A. K. Baiova: 'Istoriia voennago iskusstva kak nauka,' " *Izvestiia imperatorskii Nikolaevskoi voennoi akademi 15* (1911). Cited in Zhilin, *Russkaia voennaia mysl' konets XIX–nachalo XX v,* 48. See also A. Ageev, "Ofisery russkogo general'nogo shtaba ob opyte russko-iaponskoi voiny, 1904–1905 gg.", *Voenno-istoricheskyi zhurnal* 8 (1975): 99; Avdeev, "Voenno-istoricheskie issledovaniia v akademii general'nogo shtaba russkoi armii," *Voenno-istoricheskyi zhurnal* 12 (1987): 79.

30. Ageev, "Ofisery russkogo general'nogo shtaba ob opyte russko-iaponskoi voiny, 1904–1905 gg.," 99.

31. Ibid., 100.

32. Ibid.

33. Rostunov, *Russkaia front pervoi mirovoi voiny,* 63: "The general plan of state defence should be constructed concisely and clearly for one or two decades. In its confirmation it must be undeviating and directly translatable into practice."

34. Ageev, "Ofisery russkogo general'nogo shtaba ob opyte russko-iaponskoi voiny, 1904–1905 gg," 103.

35. Ibid., 101.

36. Carton 7N 1477, Moulin, "Appréciation du Général Martynoff sur Kouropatkin" (Dec. 7, 1906).

37. Carton 7N 1477, Moulin, "Critiques du Général Martynoff" (Dec. 16, 1906).

38. Carton 7N 1477, Moulin, "Quelques réflexions de Dragomiroff sur Kouropatkin" (Dec. 26, 1906).

39. "Sluzhba v Glavnom shtabie i Glavnom upravlenii general'nogo shtaba (1903–1908 gg.)," E. A. Nikol'skii Collection, Hoover Institution on War, Revolution, and Peace, 12.

40. Carton 7N 1477, Moulin, "Suite des 'Enseignements de la Guerre de Mandchourie' par le Lt. Colonel Nieznamoff" (Jan. 2, 1905).

41. Ageev, "Ofisery russkogo general'nogo shtaba ob opyte russko-iaponskoi voiny, 1904–1905 gg.," 103.

42. Ibid., 102.

43. Ibid., 100. The main example of this educational obsolescence and ignorance of emerging military technologies was General Linievich. See Menning, "Bayonets before Bullets," 163.

44. Ibid., 103.

45. Carton 7N 1477, Moulin, "Critiques du Général Martynoff" (Dec. 16, 1906). For a recent study on this remarkable man's personality and contributions to Russian military-theoretical development, see M. S. Martynova-Savchenko, "Nashe nasledie" and E. I. Martynov's brochure *Za chto ia byl predan sysdy i osuzhden*, reproduced in *Voennyi-istoricheskii zhurnal* 8 (1989).

46. Ageev, "Ofisery russkogo general'nogo shtaba ob opyte russko-iaponskoi voiny, 1904–1905 gg.," 103.

47. *Comptes rendus, publiés par le Rousski Invalide de conférénces sur la guerre russo-japonaise faites à l'Académie d'État-Major Nicholas* (Paris: Charles-Lavauzelle, n.d.), 63.

48. Ibid., 66.

49. Ibid.

50. A. V. Gerua, *Poslie voiny: O nashe armii* (St. Petersburg, 1907), 8.

51. Ibid.

52. Gerua based his recommendations for the change in attitude required to successfully fight modern war on the works of Blum, a famous theorist at the Berlin Military Academy in the 1880s and 1890s. Arguing that theory and routine were both extreme aspects of the military profession, Blume was to emphasize praxis—a healthy compromise between the two (a thesis which was to form the basis of the argument for a unified military doctrine).

53. Ibid., 32.

54. Ibid., 16–18.

55. Ibid., 19.

56. Ibid., 20.

57. Ibid., 34.

58. Ibid., 35, 40.

59. Ibid., 37.

60. N. Mikhnevich, *Strategiia,* Kniga I, izdanie 3e, ispravlennoe (St. Petersburg, 1911), 4.

61. Ibid., 19.

62. Ibid., 22.

63. Ibid., 10.

64. Ibid., 13.

65. N. Mikhnevich, *Osnovy strategii* (St. Petersburg, 1913), in L. G. Beskrovnyi, pod red., *Russkaia voenno-teoreticheskaia mysl' XIX i nachala XX vekov,* 455.

66. Ibid., 461.

67. Ibid.

68. Ibid., 459.

69. Ibid., 466.

70. At the time that Mikhnevich was giving this lecture, the "Great Military Program" was in the process of being ratified by the Duma. From this comment one can assume that Mikhnevich fully supported the enormous retooling of the Russian army instigated by War Minister Sukhomlinov.

71. Ibid., 461: "One needs the interference of the state in many aspects of the nation's society in wartime so that starvation and catastrophe will not occur from the reduction of moving stock on domestic railways; it is necessary to prepare a plan of economic regulation to supply all the needs of the country's population."

72. A. A. Neznamov, *Sovremennaia voina. Deistviia polevoi armii* (St. Petersburg, 1912), 2.

73. General Lewal was the first commander of the new *École Supérieure de Guerre* established in 1880 in Paris on the model of the Prussian Military Academy (*Kriegsakademie*) in Berlin. This new system emphasized the advantages of an "open" corps of staff officers recruited from all branches of the services as opposed to the previous French "closed" system which promoted an élite general staff characterized by excessive bureaucratism. Lewal was a formidable military theorist in his own right who argued that the highly rational military system which had formed and supported the military virtues of Moltke the Elder were not personal or uniquely national, but universal, and therefore adaptable by other military systems. For further discussion of his work and thought, see A. Mitchell, " 'A Situation of Inferiority': French Military Reorganization after the Defeat of 1870," *American Historical Review* 86 (1981); and General Lewal, *Études de guerre. Tactique de Marche* (Paris, 1876).

74. Ibid., 8–9.

75. Ibid., 9.

76. Ibid., 19–20.

77. Rostunov, *Russkaia front pervoi mirovoi voiny,* 88; Carton 7N 1477, Moulin, "Situation de l'armée russe dans l'hypothèse d'un conflit avec l'Allemagne" (June 27, 1905). This was the attitude of General Polivanov, quartermaster general on the main staff immediately before the creation of the main administration of the general staff.

78. Kavtaradze, "Iz istorii russkogo general'nogo shtaba," 91.

79. A. Neznamov, *Plan voiny* (St. Petersburg, 1913), in L. G. Beskrovnyi, pod red., *Russkaia voenno-teoreticheskaia mysl' XIX i nachala XX vekov,* 677.

80. Mikhnevich, *Osnovy strategii,* 527. See also I. Maslov, "Osnovnaia ideia plana voiny (strategicheskoe izsledovanie)," *Voennyi sbornik* 11 (1905): 78–80.

81. Carton 7N 1485, letter from Captain Janin to General Vignal, Aug. 4, 1911.

82. Neznamov, *Plan voiny,* 677–78. See also Maslov, "Osnovnaia ideia plana voiny (strategicheskoe izsledovanie)," 84.

83. Ibid., 679; Mikhnevich, *Osnovy strategii,* 497.

84. Maslov, "Osnovnaia ideia plana voiny (strategicheskoi izsledovanie)," 87.

85. Neznamov, *Plan voiny,* 673.

86. Mikhnevich, *Osnovy strategii,* 487.

87. Ibid., 489; Neznamov, *Plan voiny,* 689.

88. Neznamov, *Plan voiny,* 684; and *Sovremennaia voina,* 26.

89. J. W. Kipp, *From Foresight to Forecasting: The Russian and Soviet Military Experience* (College Station, Tex.: Texas A and M University, 1988), 39. Kipp cites the articles of Izmest'ev who sought to better understand the conduct of operations in the First World War through the manipulation of statistical information ("Znachenie raschete pri razrabotke i pri vedenii voennykh operatsii," *Voennyi sbornik* 3 [1915]; and "Znachenie razcheta pri razrabotke i vedenii voennykh operatsii," *Voennyi sbornik* 4 [1916]). P. I. Izmest'ev probably also wrote a work entitled *Sovremenniia tendentsii v strategii i taktik po opytu poslednei voiny,* izd. 2-e (St. Petersburg, 1910).

90. Neznamov, *Sovremennaia voina,* 11.

91. Mikhnevich, *Osnovy strategii,* 478.

92. Neznamov, *Sovremennaia voina,* 8–9. See quote on page 289.

93. Mikhnevich, *Osnovy strategii,* 481. Mikhnevich discussed the various responsibilities of this central command organ and referred to the "Position on Field Administration of Armies in War-Time."

94. Neznamov, *Sovremennaia voina,* 16–17.

95. Mikhnevich, *Osnovy strategii,* 500–502.

96. Neznamov, *Sovremennaia voina,* 18. B. Koliubakhin's article, "Element sluchainostei na voine," *Voennyi sbornik* 9 (1906), provides an interesting discussion of integrating the element of chance into the existing theory of military art.

97. Ibid., 36–37.

98. Ibid., 137–38.

99. Ibid., 65: "the faster we can move the less time the enemy has to plan and the more the information will answer [the form of] the actions." See also Mikhnevich, *Osnovy strategii,* 501.

100. Ibid., 67.

101. Ibid., 71.

102. Mikhnevich, *Osnovy strategii,* 519. See n. 73 for a discussion of Lewal.

103. N. N. Golovin, *Sluzhba general'nogo shtaba* (St. Petersburg, 1912), 1.

104. Mikhnevich, *Osnovy strategii,* 480–81; Neznamov, *Sovremennaia voina,* 8.

105. Golovin, *Sluzhba general'nogo shtaba,* 63.

106. Ibid., 64.

107. Ibid., 16.

108. Neznamov, *Sovremennaia voina,* 8–9.

109. For a more detailed discussion of Neznamov's contributions to tactical theory, see A. Ageev, "Voenno-teoreticheskoe nasledie A. A. Neznamova," *Voenno-istoricheskyi zhurnal* 11 (1983), which cites his following works: *Iz opyta russko-iaponskoi voiny (Zametki ofitsera General'nogo shtaba)* (St. Petersburg, 1906); *Voennaia istoriia. Russko-iaponskaia voina 1904–1905 gg. Operatsiia na r. Shakhe [Sha-ho]. Lektsii General'nogo shtaba podpolkovnika A. A. Neznamova v 1907–1908 uchebnom godu* (St. Petersburg, 1908); *Elementy ukrepleniia polevykh pozitsii* (St. Petersburg, 1904); *Tekushchie voennye voprosy. Voprosy sovremennoi taktiki* (St. Petersburg, 1909); and *Boi (Etiud po prikladnoi taktike)* (St. Petersburg, 1912).

110. Neznamov, *Sovremennaia voina,* 25.

111. Mikhnevich, *Osnovy strategii,* 527.

112. Maslov, "Osnovnaia ideia plana voiny (strategicheskoe izsledovanie)," 79–84.

113. Carton 7N 1474, Moulin, "Suites des 'Enseignements de la Guerre de Mandchourie' par le Lt. Colonel Nieznamoff" (Jan. 2, 1905).

114. Menning, "Bayonets before Bullets," 479.

115. Carton 7N 1477, Moulin, "Suites des 'Enseignements de la Guerre de Mandchourie' par le Lt. Colonel Nieznamoff" (Jan. 2, 1905).

116. W. T. Wilfong, "Rebuilding the Russian Army, 1905–1914: The Question of a Comprehensive Plan for National Defense" (Ph.D. diss., Indiana University, 1977), 94: "The Council of Imperial Defense . . . has been abolished by an Imperial rescript, that states that it had been found necessary to modify the manner and conditions in which the unity of action of the higher commands of the army and navy, and their co-operation with other administrations, are secured."

117. Ibid., 148–49.

118. Rostunov discusses this plan in his chapter on strategic planning before the First World War (*Russkaia front pervoi mirovoi voiny,* 91–92).

119. Wilfong, "Rebuilding the Russian Army," 142. See also Carton 7N 1485, Lt.-Col. Janin, "Rapport sommaire du Lieut. Colonel d'Infanterie brévété Janin à l'issue de sa mission à l'Académie de Guerre Nicolas à St. Petersbourg, 1910–1911" (Oct. 1911), 3, for a summary of the dual strategic posture to be pursued on the Russo-German-Austro front. This was not to contradict the official French strategic planning of General H. Bonnal and M. J. Miribel who planned for such a defensive-offensive posture in Plans 14 (1898) through 16 (1909). For a discussion of this tradition, see J. M. House, "The Decisive Attack: A New Look at French Infantry Tactics on the Eve of World War I," *Military Affairs* (Dec. 1976).

Janin also mentions in his reports to the second bureau of the French general staff (Janin, letter to Vignal, Feb. 3, 1911) that there was a "special project" involving Danilov and some of the professors of the Imperial Nicholas Military Academy, but gives no more details. This suggests the possibility of the academy's involvement in the compilation of these new mobilization plans (Plan No. 19).

120. Wilfong, "Rebuilding the Russian Army," 162.

121. General Shcherbatchev replaced General Mikhnevich at the Imperial Nicholas Military Academy in 1909.

122. Carton 7N 1485, Janin, letter to Vignal, 14/27 Oct. 1911, no. 1. In his conversation with the director of the academy, Shcherbatchev, Janin praised the work done by the professors at the academy. At this time the staff consisted of Danilov, who besides attending to his duties on the main staff was a professor of military history; General Ianushkevich, professor of military administration; General Elchaninov, professor of strategy; Colonel Neznamov, professor of tactics; and Colonel Golovin. Others were Lieutenant-General Koliubakin, Major-General Gulevich, Major-General Christiani, Colonels Baiov, Keltcherskii, Medvedev, Sergievskii, Dobruchin, and Bonch-Bruevich (from the list of recommendations for invitations and decorations at the end of Janin's *Sommaire*).

123. Carton 7N 1485, Janin, letter to Vignal, 27/10, 1910. In his *Sommaire* Janin was to observe that these doctrines were, of course, to be analogous with French military doctrines.

124. V. Domanevskii, "Edinstvo voennoi doktriny i polevoi ustav," *Russkyi invalid* 5 (1912).

125. Ibid., emphasis added.

126. Ibid.: "Therefore, military doctrine is the code of guiding views accepted in a given army at a given epoch . . . to promote mutual confidence, to create an atmosphere of spiritual independence, to promote initiative."

127. Ibid.

128. Zhilin, *Russkaia voennaia mysl' konets XIX–nachalo XX v,* 144.

129. Ibid., 145.

130. Ibid.

131. Ibid., 146.

132. Carton 7N 1487, Colonel Matton, "Note sur les invitations faites par le Gouvernment Francaise à des Officiers Russes ou inversement," (May 12, 1912).

133. Carton 7N 1478, Wehrlin, "Exercises tactiques à l'Académie de Guerre Nicolas" (Feb. 1, 1914).

134. Carton 7N 1478, Laguiche, "Le Général Engalitcheff, commandant l'Académie Nicolas" (Mar. 27, 1914).

Epilogue

The outbreak of war between the Russian Empire and the Central Powers in August 1914 brought an end to more than seventy-five years of military-theoretical development and the search for a unified military doctrine at the Nicholas Academy of the General Staff. At the heart of this military-academic endeavor there had existed a fundamental dialectic between the subjective and objective components of Russian military theory which had reached an institutional compromise in the *Field Regulations* of 1912. When world war, revolution, and civil strife eventually subsided this military-theoretical tradition continued to play an integral part in the development of the Soviet Red Army.

Whether through the dictates of circumstance or out of a sense of duty to the preservation of Russia from its external enemies many officers of the rapidly disintegrating Imperial Army in the months after the October Revolution found themselves responsible for the defense of the Petrograd Soviet. A significant number of these officers were graduates and former professors from the Nicholas Academy, including such visionaries of Imperial military theory as N. P. Mikhnevich, A. A. Neznamov, V. F. Novitskii, A. M. Zaionchkovskii, M. I. Bonch-Bruevich, P. I. Izmest'ev, and A. A. Svechin who, as military specialists (*voenspetsi*), were to become influential contributors to the birth and development of Soviet military doctrine.[1]

Throughout the 1920s these military specialists in conjunction with their revolutionary counterparts, M. N. Tukhachevskii, M. V. Frunze, V. K. Triandafillov and others, reviewed this theoretical legacy in terms of the ideological requirements of the new socialist state and the military-operational experiences gained in six years of war. Frunze, in particular, was concerned with molding this legacy with the sociological laws of Marxism-Leninism in an attempt to equip the Red Army with a uniquely socialist unified military doctrine:

> Of course, one must not speak about absolute permanence and stability. There is nothing absolute and permanent or unchanged. This is the first premise of our Marxist theory. We approach an examination of all phenomena surrounding us from this viewpoint. For this reason, without having in mind the creation of any permanent, ossified . . . ideas, we should still seek homogeneous phenomena in the chaos surrounding us, generalize them into a single whole and endeavor to analyze the phenomena surrounding us, otherwise we will be lost and non-comprehending when confronted by them and rush into the area of harmful practical empiricism. The question of the need to seek out certain generalizing aspects is a question that is fully resolvable, legitimate and natural and the resolving of it will help us come closer to a resolution of military problems.[2]

Within the context of this socially objective theoretical tradition the ultimate goal, prediction (in the form of operational foresight and long-range forecasting of military-technological innovation) gained renewed impetus.[3]

This ease of doctrinal transition from Imperial to Red Army was facilitated, in part, by the common philosophical basis that the critical historical method shared with dialectical materialism. In order to distinguish those parts of Soviet military doctrine that were borrowed from the Imperial legacy from those which were the unique contributions of Marxism-Leninism one must turn to the postwar development of theory in the Russian military emigré community.

After the Russian Civil War drew to a close in the Autumn of 1920 fragments of the White Russian Army went into exile to eventually join the existing emigré communities led by Grand Duke Nikolai Nikolaevich in Paris and General Wrangel in Belgrade. Ever hopeful of regaining control of Russia from the Bolsheviks, these two military figures sought to reorganize the remaining forces loyal to the nationalist cause and train an élite officer corps in the professional ethic of the old General Staff as well as in the latest military-technological developments indicated by the experience of the First World War. They called upon the expertise of General Golovin, a former professor at the Nicholas Academy, to formulate a doctrine for the new army and to supervise the re-establishment of higher military education.

Upon evaluating the combat experience of the Russian Imperial Army in the world war General Golovin submitted a report to the Grand Duke in 1926 entitled "Thoughts on the Future Construction of the Russian Armed Forces," which blamed the defeat of the Russian army on the incompetency of the high command and a low level of technical training in an officer corps unfamiliar with the demands of modern warfare.[4] Referring back to the argument for an implicit and explicit *edinonachale* raised after the Russo-Japanese debacle and the soundness of Mikhnevich's and Neznamov's theories on war and combat Golovin suggested a training program founded on the educational principles of the former Nicholas Academy. A year later he formally established the higher military-scientific courses in Paris and Belgrade on the basis of an extensive curriculum of theory and the old method of teaching tactics through wargames.

In the mid-1930s Golovin summarized the entire Russian military-theoretical tradition in a speech entitled "The Sociology of War," which he gave to an international gathering of sociologists and later published as a text for his students.[5] In this speech he argued that there was a need to recognize the division of contemporary military science into two branches. The first, and most general, was the science (or sociology) of war in the context of the positivist social sciences, while the more specifically martial science was the science of conducting warfare, traditionally known as the "theory of military art." Responsible for investigating the very general laws of correlation between natural and social phenomenon, the sociology of war was to indicate the margins of reliability in which specific principles of conducting warfare would successfully operate. In turn, these specific principles were to provide a fundamental concept, or intellectual *edinonachale,* which would regulate creative effort on the battlefield. A constantly re-evaluated synthesis of these laws and principles would provide contemporaneity and avoid scholasticism in military science. The ultimate goal of this sociology of war predicated on the gradual accumulation of empirical information and the systematization of analytical methodologies was the pursuit of *prediction:*

> The degree of possibility of attaining this last stage varies in different sciences. The more complicated and numerous are the elements of the phenomena under investigation, the greater is the requisite knowledge of contributory circumstances that is essential in order to arrive at an exact prediction. The science of war, as all social sciences, has to deal with phenomenon of endless complexity depending on the multiplicity of contributing circumstances and their varying quality. The complexity of the phenomenon under investigation is so great that it will never be possible to decompose a given phenomenon into its component elements; this being necessary in the process of arriving at an exact prediction when the only quantity remaining unknown is the sought-for consequence; owing to the complexity of the phenomena an element of vagueness will remain in the contributing causes, and in this case even mathematics arrive at a whole series of solutions (indefinite equations). Therefore, even after attaining perfect development the science of war will only reach the last stage (prediction) conditionally, and will chiefly remain in the stage of investigation more than any other science.[6]

Although Golovin attempted throughout the 1930s to persuade various national military establishments to adopt this military-scientific tradition his arguments ultimately remained within the confines of the Russian military emigré community in the two cities and suffered from the lack of stimulation an institutionally active army would have normally provided. Even his attempts to augment the lessons of the wargames with field trips had to be abandoned when these activities came to the attention of the local French police. When the formal courses were discontinued in January 1944 this academic hybrid slipped into obscurity.[7]

Despite the new priorities in securing the gains of the October Revolution, advances in military technology, and the emergence of new theories and vocabulary to integrate all these changes, the basic tenets of Marxism-Leninism in military affairs did not differ substantially from the interrelated, nineteenth-century concepts of numeralization, materialization, and democratization. Given the current era of military-institutional reform and theoretical debate over changing responsibilities in Soviet military doctrine, this review should help in providing insight into the philosophical syntax of concepts involved in these contemporary military issues.

NOTES

1. The military-theoretical legacy of the Russian imperial army in the origins and development of the Soviet Red Army is the subject of several Soviet studies, the most recent and important one being A. G. Kavtaradze's *Voennye spetsialisty na sluzhbe Respubliki Sovetov 1917–1920 gg.* (Moscow: Nauka, 1988). Other recent studies are: S. A. Fediukin, *Sovetskaia vlast' i burzhauznye spetsialisty* (Moscow: Mysl, 1965); I. A. Korotkov, *Istoriia sovetskoi voennoi mysli 1917–1941* (Moscow, 1980); and P. A. Zhilin, pod red., *Zarozhdenie i razvitie sovetskoi voennoi istoriografii, 1917–1941* (Moscow: Nauka, 1985). Western analysis of this philosophical and institutional continuum can be found in M. Mayzel's *Generals and Revolutionaries. The Russian General Staff during the Revolution: A Study in the Transformation of Military Elite* (Osnabruck: Biblio Verlag, Studien zur Militargeschichte, Militarwissenschaft und Konflictforschung, 1979) as well as J. Erickson's *The Russian Imperial/Soviet General Staff* (College Station, Tex.: Texas A & M University, 1984); and "The General Staff: Theory and Practice from the Tsarist to the Soviet Regime," in *National Security Policy: The Decision-Making Process,* edited by A. L. Pfaltzgraff, Jr., and U. Ra'anan, pp. 15–85 (Medford, Mass.: Archon, 1984).
2. M. A. Gareev, *M. V. Frunze—Military Theorist* (London: Pergamon-Brassey's, 1988), 104.
3. At the heart of this concept lies a distinction between the theory of contemporary war and future war. In Soviet military-political parlance it is merely a temporal periodization but I would offer a military-theoretical distinction. Whereas the concept of contemporary war is a function of the theory of military art in its subjective (Clausewitzian) sense the concept of future war is a function of that theory in its objective (Jominian) sense, or more appropriately, the theory of strategic planning.
4. N. N. Golovin Collection, box no. 15, "Russian Army, Report on the Future," Hoover Institution on War, Revolution, and Peace, 8–14.
5. Only a Russian language edition was published: *Nauka o voine: o sotsiologicheskom izuchenii voiny* (The Science of War: On a Sociological Study of Warfare) (Paris: Signal, 1938).
6. N. N. Golovin Collection, box no. 11, "Sociology of War," Hoover Institution on War, Revolution, and Peace, 33.
7. John L. Bates Collection, "Zarubezhnye vysshie voenno-nauchnye kursy pod rukovodstvom professors General-Leitenanta N. N. Golovin. 1927-Marta 22–1977." Hoover Institution on War, Revolution, and Peace.

Bibliography

UNPUBLISHED PRIMARY SOURCES

Ministry of Defense of the French Republic

Service Historique de l'Armée de Terre. Vincennes, Paris. *Archives de la Guerre: E. M. A. et Attachés Militaires, Russie.* Carton 7N 1465–1535 (1872–1914).

Carton 7N 1465

"Armée Russe. Renseignements sur le Corps d'État-major" (Nov. 1872).
General Le Flô, French ambassador to Russia. "Opinions russes sur le procès Bazaine" (Dec. 25, 1873).
Colonel Gaillard. "Sur la liberté de discussion en matière technique dans l'armée russe" (Sept. 10, 1874).
______. "Renseignements divers" (Sept. 14, 1874).

Carton 7N 1467

Lt.-Colonel de Boisdeffre. "Le Général Dragomiroff et l'école d'État-major" (Dec. 7, 1880).

Carton 7N 1469

Lt.-Colonel de Sermet. "Échanges de documents proposés par le Général Dragomiroff entre les deux écoles superiéures de guerre" (Nov. 9, 1884; Feb. 10, 1885).

Carton 7N 1470

Captain Moulin. "Forces russes sur l'éventuels theatre d'opérations" (May 1, 1886).

______. "Plan de concentration en cas de guerre contre l'Allemagne et l'Autriche" (June 12, 1886).
(Description of the Russian Mission to be sent to observe large maneuvers in France) (June 19, 1886).
______. "Les manoeuvres 'de bataillon' proposées par le Général Dragomiroff" (Oct. 25, 1886).

Carton 7N 1471

Moulin. "Opinion du Général Leer sur l'état de preparation à la guerre du l'armée russe" (June 6, 1887).
______. "Commandement des armées en temps de guerre" (June 9, 1887).
Colonel de Briois. "Constitution d'une section historique à l'État-major général de l'armée russe" (Feb. 24, 1888).
Moulin. "Nouvelles nominations dans l'armée" (Sept. 7, 1889).
______. "Opinion du Général Stroukoff sur les raids en Prusse Orientale" (Oct. 10, 1889).
______. "Envoi du plan de concentration no. 1 pour 1889" (Oct. 10, 1889).
______. "Incident au départ du Général Dragomiroff pour Kief" (Oct. 27, 1889).
______. "Au sujet des manoeuvres d'automne du Général Gourko sur la Narew" (Nov. 12, 1889).

Carton 7N 1472

Moulin. "Rapport sur les manoeuvres de l'armée russe en Volkynie" (Nov. 15, 1890).
______. "Au sujet d'un cours pratique pour les officiers du Cavalerie en Pologne" (June 14, 1892).
______. "Reunion des chefs d'État-Major de Varsovie, Vilna et Kief à Petersbourg" (Feb. 1, 1895).

Carton 7N 1473

Moulin. "Conversation de Dragomiroff avec l'Empereur" (Mar. 15, 1893).
______. "Opinion du Général Pouzyrevsky et du Général Dragomiroff sur les chances du sucès de la Russie dans une guerre prochaine" (Apr. 5, 1893).
______. "Conversation avec un général russe au sujet du plan d'opération de l'armée de Gourko" (Dec. 22, 1893).
______. "Sur quelques études de tactique faits dans la Circonscription de Varsovie" (Oct. 24, 1894).
______. "Sur la revision du règlement d'infanterie" (Dec. 8, 1894).
______. "Conversations avec le Général Dragomiroff" (July 28, 1895).

Carton 7N 1474

Moulin. "Réunion d'une commission presidée par Gourko" (Jan. 30, 1896).
______. "Sur les projects d'operations du Général Dragomiroff" (Jan. 23, 1897).

Carton 7N 1475

Moulin. "Sur le nouveau Ministre de la Guerre et le nouveau chef de l'État-major" (Jan. 16, 1898).

French ambassador in Russia to the minister of foreign affairs. "La retraite des généraux Vannovsky et Obroutcheff" (Jan. 19, 1898).
Moulin. "Nouvelles nominations dans l'armée" (Feb. 10, 1898).
_____. "Changements dans le haut personnel de l'armée russe" (Feb. 25, 1898).
_____. "Nouvelles nominations de gouverneurs généraux" (Apr. 10, 1898).
_____. "Sur le Général Kouropatkine" (Oct. 11, 1898).
_____. "Au sujet du double courant d'opinion de l'État-major russe a propos du plan de concentration en cas de guerre avec la Triple Alliance" (Apr. 16, 1899).

Carton 7N 1477

Moulin. "Suite des 'Enseignements de la Guerre de Mandchourie' par le Lt. Colonel Nieznamoff" (Jan. 2, 1905).
_____. "Situation général de la guerre" (Feb. 10, 1905).
_____. "Sur l'arrivée à Petersbourg du Général Dragomiroff" (Mar. 10, 1905).
_____. "Sur le remplacement du Général Kouropatkine" (Mar. 25, 1905).
_____. "Situation de l'armée russe dans l'hypothèse d'un conflit avec l'Allemagne" (June 27, 1905).
_____. "Réorganisation du haut commandment de l'armée" (July 10, 1905).
_____. "Sur le Général Dragomiroff" (Nov. 1, 1905).
_____. "Au sujet du rôle des membres de la famille impériale dans l'armée" (Nov. 20, 1906).
_____. "Appréciation du Général Martynoff sur Kouropatkin" (Dec. 7, 1906).
_____. "Critiques du Général Martynoff" (Dec. 16, 1906).
_____. "Quelques réflexions de Dragomiroff sur Kouropatkin" (Dec. 26, 1906).

Carton 7N 1478

Wehrlin. "Exercises tactiques à l'Académie de Guerre Nicolas" (Feb. 1, 1914).
Laguiche. "Le Général Engalitcheff, commandant l'Académie Nicolas" (Mar. 27, 1914).
_____. "Conversations avec le Commandent de l'Académie Nicolas" (Apr. 24, 1914).

Carton 7N 1480

M. Bouard, French council general in Warsaw to the minister of foreign affairs. "La guerre future et ses consequences économiques" (Apr. 24, 1893).

Carton 7N 1484

Captain P. Dreyfus. "Étude sur l'armée russe. Le Général Dragomirof" (Perignan, May 26, 1896).
A. Nessel. "L'Académie d'État-major Nicolas à St. Petersbourg" (Paris, Feb. 26, 1897).

Carton 7N 1485

Captain Mahon. "Rapport sur un séjour à Varsovie. Section III, Rapports particulars avec les officiers de l'entourage du Général Pouzerevski" (Oct. 1902).
Colonel Matton. "Projects du Ministre de la Guerre. Augmentation des effectifs. Proposition d'envoyer un officer stagiaire a l'Académie d'État-major" (Nov. 1, 1910).
Janin a General Vignal. Letters from Oct. 14, 1910, to Aug. 18, 1911.

Lt.-Colonel Janin. "Rapport sommaire du Lieut. Colonel d'Infanterie brevété Janin à l'issue de sa mission à l'Académie de Guerre Nicolas à St. Petersbourg, 1910–1911" (Oct. 1911).

Carton 7N 1487

Colonel Matton. "Note sur les invitations faites par le Gouvernment Francaise à des Officiers Russes ou inversement" (May 12, 1912).

Carton 7N 1535

"Notice statistique sur l'armée russe: La physionomie de l'armée russe. Les Doctrines" (1913).

The British Library, Department of Manuscripts (London)

Egerton MSS. 3166, ff. 91–93, 112, 115, 122, 126. General A. H. Jomini's letters to Tsar Alexander II and War Minister General D. A. Miliutin with regard to reform of the Russian military-educational system, 1864.
Egerton MSS. 3168, ff. 36–64, 67. Letters in response from Alexander II and D. A. Miliutin, 1864.

The National Library of Scotland, Manuscripts Room (Edinburgh)

MSS. 13088, ff. 34–34. Correspondence of Colonel, the Honorable Frederick Arthur Wellesley, British military attaché in St. Petersburg with regard to mobilization of the Russian army for the Russo-Turkish War, 1877/78.

Hoover Institution on War, Revolution, and Peace (Stanford, Calif.)

Russia, Voennyi agent (France), Records

Box no. 4. Folio. Correspondence between the main administration of the general staff (GUGSh) in St. Petersburg and the Russian military attaché in Paris, 1864. Document nos. 4431, 7917, 8716, 188.
Box no. 5. Folio. Correspondence between the main administration of the general staff (GUGSh) in St. Petersburg and the Russian military attaché in Paris, 1865. Document nos. 1330, 33, 5432, 5198, 5549.

E. A. Nikol'skii Collection

"Sluzhba v Glavnom shtabie i Glavnom upravlenii general'nogo shtaba (1903–1908 gg.)."

Nicholai N. Golovin Collection

Box no. 15. Subject file, 1914–1943. "Mysli ob ustroistve budushchei Rossiiskoi vooruzhennoi sily" (Russian army report on future organization to Grand Duke Nicholai Nicholaevich, 1926?).
Box no. 11. "Sociology of War."

John L. Bates Collection

"Zarubezhnye vysshie voenno-nauchnye kursy pod rukovodstvom professora General-Leitenanta N. N. Golovin. 1927-March 22–1977."

PUBLISHED PRIMARY SOURCES

Russian Collections, Official Histories, and Imperial Military Encyclopedias

Entsiklopedicheskii slovar. Izd. F. A. Brokgauz, I. A. Efron. St. Petersburg, 1891–1904.

Avstro-Vengriia. Voenno-statisticheskoe opisanie. Chast'I. Vostochno-galitsiuskii raion'. Pod obshchei redaktsiei general'nogo shtaba polkovnika Samoilo, sostavila general'nogo shtaba polkovnik' Pototskii. St. Petersburg: Glavnoe upravlenie general'nogo shtaba, Voennaia Tip. v zdanii Glavnogo Shtaba, 1912.

Entsiklopediia voennykh i morskikh nauk. Pod red. G. A. Leer, Tom. 1, 3, 5, 7. St. Petersburg, 1883–1897.

Entsiklopedicheskogo slovariia. Pod red. A. i I. Granat. St. Petersburg, 1910–1938.

Istoriia russkoi armii i flot. Tom. 1-2 Moscow, 1913.

Materialy dlia opisaniia russko-turetskoi voiny 1877–1878 gg. na Kavkazsko-Maloaziatskom teatr. Tom. 1. St. Peterburg, 1904.

Opis del sekretnogo pobyt'iia moskovskogo otdeleniia obshchago arkhiva glavnogo shtaba. (47-ia opis). Pod red. i s predisloviem general'nogo shtaba polkovnika Maslovskogo. St. Petersburg, 1890.

Osoboe pribavlenie k opisaniiu russko-turetskoi voiny 1877–1878 gg. na Balkanskom poluostrov. Vypusk IV. Soobrazheniia, kasaiushchiiacia plana voiny. St. Petersburg, 1901.

Russko-Iaponskaia voina 1904–1905 gg: Rabota Voenno-istoricheskoi kommissii po opisaniiu russko-iaponskoi voiny. Tom. 1-9. St. Petersburg, 1910–1913.

Sbornik sistematicheskikh soobshchenii po istorii russko-iaponskoi voiny, sdelannikh v Vilensko, voennom sobranii v techenie zimny perioda 1907–1908 gg. Chast' 1–2. Vilna, 1907/8.

Sbornik voenno-istoricheskikh materialov. St. Peterburg, 1892–1904.

Stoletiie voennogo ministerstva, 1802–1902. Pod red. D. Skalon, Tom. 1-13. St. Petersburg, 1902–1914.

Voennaia entsiklopediia. Pod red. V. I. Novitskii, Tom 1-8. St. Petersburg, 1911-1915.

"Voennaia reforma." Sbornik statei izdanie *Voennogo Golosa.* St. Petersburg, 1906.

Voenno-statisticheskoi sbornik. Pod red. N. N. Obrucheva, Tom. 1-4. St. Petersburg, 1867-1871.

Voennyi entsiklopedicheskyi leksikon. Pod red. L. I. Zeddeler, Tom. 1. St. Petersburg, 1837-1849.

Russkaia voenno-teoreticheskaia mysl' XIX i nachala XX vekov. Pod red. L. G. Beskrovnogo. Moscow: Voennoe Izdatel'stvo Ministerstva Oborony Soiuza SSR, 1960). The following works are brought together in this collection:

N. Medem. *Obozrenie izvestneishikh pravil i sistem strategii.* St. Petersburg, 1836.

______. *Taktika.* St. Petersburg, 1837.

P. A. Yazykov. *Opyt teorii strategii.* St. Petersburg, 1842.

D. Miliutin. *Pervye opyty voennoi statistiki.* St. Petersburg, 1842.

_____. *Iz stat'e "Starcheskie razmyshleniia o sovremennom polozhenii voennogo dela v Rossii."* St. Petersburg, 1909; repr. 1912.

F. Goremykin. *Rukovodstvo k izucheniiu taktiki. Chast I. Nachal'naia taktika.* St. Petersburg, 1849.

A. Astaf'ev. *O sovremennom voennom iskusstve. Chast' I–II.* St. Petersburg, 1856–1862.

G. A. Leer. *Strategiia (taktika teatra voennikh deistvii).* St. Petersburg, 1898.

M. Dragomirov. *Uchebnik taktiki.* St. Petersburg, 1906.

A. N. Petrov. *K voprosam strategii.* St. Petersburg, 1898.

N. P. Mikhnevich. *Vliianie noveishikh tekhnicheskikh izobretenii na taktiku voisk.* St. Petersburg, 1898.

_____. *Osnovy strategii.* St. Petersburg, 1913.

A. Neznamov. *Boi (etiud po prikladnoi taktike).* St. Petersburg, 1912.

_____. *Plan voiny.* St. Petersburg, 1913.

Books, Russian Imperial

Baiov, A. K. *Istoriia russkoi armii: kurs voennykh uchilishch.* St. Petersburg, 1912.

Bliokh, I. S. *Obshchie vyvody iz sochineniia "Budushchaia voina v tekhnicheskom, politicheskon i ekonomicheskom otnosheniiakh."* St. Petersburg, 1897.

Bonch-Bruevich, M. D. *Boi.* Kiev: Kievskogo voennogo okruga, 1909.

Bulla, K. K. *Bol'shie manevry v bysochaishem prisustvii pod gorodom Pskom 5–10 avgusta 1903 g.* Sobstvennikh izdaniia, fotograf. St. Petersburg, 1904.

Elizarov, N. S. *Nashi sosiedi i niekotoriia drugiia glavnieishiia gosudarstva: obzor ikh voennoi moshchi.* Moscow: Okruzhnogo shtaba, 1913.

Geisman, P. *Kratkii kurs istorii voennogo iskusstva v srednie i novye veka.* St. Petersburg, 1893–1896.

_____. *Opyt issledovaniia taktiki massovykh armii.* St. Petersburg, 1894.

Gerua, A. V. *Poslie voiny: O nashe armii.* St. Petersburg, 1907.

Glinoetskii, N. P. *Istoricheskii ocherk Nikolaevskoi akademii general'nogo shtaba.* St. Petersburg, 1882.

Golovin, N. N. *Sluzhba general'nogo shtaba.* St. Petersburg, 1912.

Grigor'eyev, V. N. *Kak sleduet ponimat' voennuiu nauki: otviet g-nu V. Borisov na ego broshiuru "Strategicheskie voprosy" ili razbor polozhenii sochineneii Generala Leer "Strategiia".* St. Petersburg, 1904.

Konspekt strategii, sostavlen ofitserami starshogo klassa Nikolaevskoi akademii general'nogo shtaba. St. Petersburg, 1887.

Leer, G. A. *Opyt kritico-istoricheskogo issledovaniia zakonov iskusstva vedeniia voiny.* St. Petersburg, 1869.

_____. *Slozhyi operatsii.* St. Petersburg, 1892.

Martynov, E. I. *Strategiia v epokhu Napoleona I i v nashse vremiia.* St. Petersburg, 1894.

_____. *Vospominaniia o iaponskoi voinie komandira pekhotnogo polka.* Plotsk, 1910.

Men'kov, P. K. *Zapiski Petra Kononovicha Men'kova.* Tom. 1–3. St. Petersburg, 1898.

Mikhnevich, N. P. *Osnovy russkogo voennogo iskusstva. Sravnitel'nyi ocherk sostoianiia voennogo iskusstva v Rossii i Zapadnoi Evrope v vashneishiia istoricheskiia epokhi.* St. Petersburg, 1898.

_____. *Strategiia.* Kniga I, izdanie 3e, ispravlennoe. St. Petersburg, 1911.

Miliutin, D. A. *Vospominaniia,* Russian Memoir Series, 26, edited by W. B. Lincoln. Newtonville: Oriental Research Partners, 1979.

Neelov, N. *Ocherk sovremennogo sostoianiia strategii.* St. Petersburg, 1846.
Neznamov, A. A. *Sovremennaia voina. Deistviia polevoi armii.* St. Petersburg, 1912.
Novitskii, V. F. *Na puti k usovershenstvovaniiu gosudarstvennoi oborony.* St. Petersburg, 1909.
_____. *Mobilizatsiia (Teoriia mobilizatsii).* St. Petersburg, 1910.
Petrov, A. N. *Vlianie turetskikh voin s poloviny proshlogo stoletiia na razvitie russkogo voennogo iskusstva.* St. Petersburg, 1893/94.
Polievktov, M. A. *Nikolai I. Biografiia i obzor tsarstvovaniia.* St. Petersburg, 1918.
Puzyrevskii, A. K. *Razvitie postoiannikh regularnikh armii i sostoianie voennogo iskusstva v vek Liudovika XIV i Petra Velikogo.* St. Petersburg, 1889.
Sukhotin, N. N. *Zametki po predmeti istorii voennogo iskusstva.* St. Petersburg, 1881.
Ustav polevoi sluzhby: Vysochaishe utverzhden 27 apreliia 1912 g. Izd. 2e. Moscow, 1916.
Vendrikh, A. A., von. *Zavisimost' budushchikh voin ot blagoustroistva voennykh soobshchenii: soobshchenie Nikolaevskii akademii general'nogo shtaba.* St. Petersburg, 1899.
Volgin, A. M. *Ob armii.* St. Petersburg, 1908.
Yazykov, P. A. *Opyt' teorii voennoi geografii.* Tom 1–2. St. Petersburg, 1838.

Books, Non-Russian

Bloch, J. *The Future of War in Its Technical, Economic, and Political Relations.* Boston, 1902.
Caemmerer, R. von. *The Development of Strategical Science during the 19th Century.* London: Hugh Rees, 1905.
Clausewitz, C. von. *On War,* edited by P. Paret and M. Howard. Princeton: Princeton University Press, 1984.
Colin, J. *Les Transformations de la Guerre.* Paris: Flammarion, 1912.
Comptes rendus, publiés par le Rousski Invalide de conférences sur la guerre russo-japonaise faites à l'Académie d'État-Major Nicolas. Aperçu stratégique de l'offensive de la IIe armée de Mandchourie en janvier 1905. Le détachement monté du général Michtchenko pendant l'offensive de la IIe armée en janvier 1905. Paris: Charles-Lavauzelle, n.d.
Detaille, E. *Les Grandes Manoeuvres de l'Armée Russe. Souvenir du Camp de Krasnoe-Selo.* Paris: Valadon, 1886.
Dragomirov, M. I. *Principes essentiels pour la conduite de la guerre. Clausewitz interprété.* Paris: L. Boudoin, 1889.
_____. *"Guerre et Paix" de Tolstoi au point de vue militaire.* Paris: L. Boudoin, 1896.
Foch, Marshal F. *The Principles of War.* London: Chapman and Hall, 1918.
Gilbert, J. A., ed. *An Exposition of the First Principles of Grand Military Operations and Movements Compiled from the "Treatise upon Grand Military Operations" by the Baron de Jomini.* London: Whitehall, 1825.
Goltz, C. von der. *The Nation in Arms.* London: Hugh Rees, 1903.
Kuropatkin, A. N. *The Russian Army and the Japanese War: Being Historical and Critical Comments on the Military Policy and Power of Russia and on the Campaign in the Far East.* 2 vols. London, 1909.
Lauth, J. *L'État militaire des principales puissances étrangers en 1902.* Paris: Berger-Levrault, 1902.

Lloyd, H. *The History of the Late War in Germany between the King of Prussia and the Empress of Germany and Her Allies.* 2 vols. London, 1781–1790.

Mahon, P. *L'Armée Russe après la campaign de 1904–95, publié sous la direction du 2e Bureau de l'État-Major du l'Armée.* Paris, 1906.

Napoleon. Paris: Librairie Nilsson, 1894. This is a rare publication containing two unusual essays which examine the command styles of Napoleon and Moltke: one by Colmar von der Goltz ("Napoleon et Moltke") and another by M. I. Dragomirov ("Boutades Militaires").

Rapport de la Commission déléguée par ordre de S. M. l'Empereur de toutes les russies pour examiner la fabrication des pièces d'artillerie destinées aux fortresses et à la marine. Paris, 1865.

Russia in the East, 1876–1880, the Russo-Turkish War and the Kuldja Crisis as Seen Through the Letters of A. G. Jomini to N. K. Giers, edited by C. and B. Jelavich. Leiden: E. J. Brill, 1959.

Schlieffen, A. *Cannae.* Fort Leavenworth, Kans., 1936.

Story, D. *The Campaign with Kuropatkin.* London: T. Werner Laurie, 1904.

Tovey, R. E. *The Elements of Strategy.* London: Hugh Rees, 1904.

Wellesley, F. *Recollections of a Soldier-Diplomat.* London: Hutchinson, n.d.

Wilkinson, S. *The Brain of an Army: A Popular Account of the German General Staff.* London: Constable, 1895.

Articles from Imperial Russian Periodicals

N.B.: The majority of the following articles are from *Voennyi sbornik* and *Russkii invalide,* the official monthly and daily periodicals of the Russian war ministry. As the same editorial board was responsible for the publication of both periodicals, they share the same division of contents: official and unofficial articles with separate pagination. With the exception of the "Russkoe voennoe obozrenie" article in *Voennyi sbornik,* all these articles are to be found in the unofficial section of the cited periodical. As each edition of *Ruskii invalide* is only five to seven pages in length, page numbers have been omitted.

Adaridi, K. "Liubiteli doktriny." *Russkii invalid* 14 (1912).

Anon. "Russkoe voennoe obozrenie." *Voennyi sbornik* 4 (1865): 138–52.

Bonch-Bruevich, V. "Pismo v redaktsiiu." *Russkii invalid* 6 (1912).

Boretskii, A. "Ustavy i zhizn'." *Russkii invalid* 15 (1912).

Dmitrievskii. "Doktrina, dekadentstvo, shablon, doktrinerstva." *Russkii invalid* 72 (1912).

Domanevskii, V. "K novomy ustavu polevoi sluzhby." *Russkii invalid* November 19, (1911).

_____. "Edinstvo voennoi doktriny i polevoi ustav." *Russkii invalid* 5 (1912).

Dragomirov, M. I. "Otkrytoe pis'mo G. Bliokhu." *Razvedchik* 316 (1898).

Geisman, P. "Mikhail Dmitrievich Skobelev." *Voennyi sbornik* 6 (1907): 1–56.

_____. "Uchrezhdenie Imperatorskoi voennoi akademii." *Voennyi sbornik* 11 (1908): 81–98; 12 (1908): 133–56.

Golitsyn, N. S. "Imperatorskaia voennaia Akademiia v 1834–48 gg." *Russkaia starina* 27 (1883): 411–14; 28 (1883): 703–12.

Gulevich, A. "Voina i narodnoe khoziaistvo." *Voennyi sbornik* 1 (1898): 60–75.

Koliubakhin, B. "Element sluchainostei na voine." *Voennyi sbornik* 9 (1906): 47–69.

Krivtsov, E. "Voennoe iskusstvo." *Russkii invalid* 16 (1912).

Leer, G. A. "Znachenie kriticheskoi voennoi istorii v izuchenii strategii i taktiki." *Voennyi sbornik* 5, (1863).

_____. "Sopostavlenie neskol'kikh evoliutsii v strategicheskom iskusstv'." *Voennyi sbornik* 3 (1893).

Maksheev, F. A. "Neskol'ko slov a Dragomirovskom i Leerovskom periodakh nachal'stvovaniia Akademiei." *Voennyi sbornik* 12 (1907): 234–238.

Maslov, I. "Osnovnaia ideia plana voiny (strategicheskoe izsledovanie)." *Voennyi sbornik* 11 (1905): 79–95.

Miliutin, D. A. "Voennye reformy imperatora Aleksandra II." *Vestnik Evropy* (1882): 5–35.

Ogorodnikov, F. "Doktrinerstvo i ustav." *Russkii invalid* 44 (1912).

Petrov, A. N. "Zadachi sovremennoi strategii." *Voennyi sbornik* 5 (1894): 35–64.

Poslie voiny: Voenno-obshchestvennyi sbornik statei. *Voennyi Golos.* St. Petersburg, 1906–1907.

Puzyrevskii, A. K. "Budushchaia voina v tendentsioznom izobrazhenii." *Razvedchik* 410 (1898).

Svechin, A. "Bol'shaia voennaia programma." *Russkaia Mysl'* (1913): 19–29.

Zaionchkovskii, A. "Edinstvo voennoi doktriny (opasnosti i uvlecheniia)." *Russkii invalid.* (December 31, 1911).

_____. "Eshche po povody doktriny." *Russkii invalid* 7 (1912).

Zuev, D. "Pismo v redaktsiiu." *Russkii invalid* 6 (1912).

SECONDARY SOURCES

Books, Soviet

Beskrovnyi, L. G. *Ocherki po istochnikovedeniu voennoi istorii Rossii.* Moscow: Izdatel'stvo Akademii Nauk SSSR, 1957.

_____. *Ocherki voennoi istoriografii Rossii.* Moscow: Izdatel'stvo Akademii Nauk SSSR, 1962.

_____. *Russkaia armiia i flot v XIX veke. Voenno-ekonomicheskii potentsial Rossii.* Moscow: Izdatel'stvo 'Nauka', 1973.

_____. *Armiia i flot Rossii v nachale XX v: Ocherki Voenno-ekonomicheskogo potentsiala.* Moscow: Izdatel'stvo 'Nauka', 1986.

Bonch-Bruevich, M. *From Tsarist General to Red Army Commander.* Moscow: Progress, 1966.

Gareev, M. A. *M. V. Frunze—Military Theorist.* London: Pergamon-Brassey's, 1988.

Istoriia otechestvennoi voenno-entsiklopedicheskie literatury. Pod red. M. M. Kir'yan. Moscow: Izdatel'stvo 'Nauka', 1980.

Kir'yan, M. M. *Problemy voennoi teorii v sovetskikh nauchno-spravochnykh izdaniiakh.* Moscow: Izdatel'stvo 'Nauka', 1985.

Makeev, N. *N. G. Chernyshevskii—redaktor "Voennogo sbornika."* Moscow: Voenizdat, 1950.

Meshcheriakov, G. P. *Russkaia voennaia mysl' v XIX v.* Moscow: Izdatel'stvo 'Nauka', 1973.

Rostunov, I. I. *Russkaia front pervoi mirovoi voiny.* Moscow: Izdatel'stvo 'Nauka', 1976.

_____ed. *Russo-turetskaia voina 1877-1878 gg.* Moscow: Izdatel'stvo 'Nauka', 1977.

Russkaia voennaia mysl' konets XIX-nachalo XX v. Pod red. P. A. Zhilin. Moscow: Izdatel'stvo 'Nauk', 1982.

Strokov, A. A. *Istoriia voennaia iskusstva.* Moscow: Voenizdat, 1965.

Zaionchkovskii, P. A. *Voennye reformy 1860-1870 godov v Rossii.* Moscow: Izdatel'stvo Moskovskogo Universiteta, 1952.

_____. *Samoderzhavie i russkaia armiia na rubezhe XIX-XX stoletii, 1881-1903.* Moscow: Izdatel-stvo 'Mysl', 1973.

Zhilin, P. A. *Istoriia voennogo iskusstva.* Moscow: Voenizdat, 1986.

Articles from Soviet Periodicals

Ageev, A. "Voenno-istoricheskie vzgliady N. P. Mikhnevicha." *Voenno-istoricheskyi zhurnal* (1975): 90-95.

_____. "Ofisery russkogo general'nogo shtaba ob opyte russko-iaponskoi voiny, 1904-1905 gg." *Voenno-istoricheskyi zhurnal* 8 (1975): 99-104.

_____. "Iz dnevnika N. P. Mikhnevicha." *Voenno-istoricheskyi zhurnal* 5 (1976): 69-75.

_____. "Voenno-teoreticheskoe nasledie A. A. Neznamova." *Voenno-istoricheskyi zhurnal* 11 (1983): 84-89.

Avdeev, A. "Voenno-istoricheskie issledovaniia v russkoi armii." *Voenno-istoricheskyi zhurnal* 3 (1986): 81-84.

_____. "Voenno-istoricheskie issledovaniia v akademii general'nogo shtaba russkoi armii." *Voenno-istoricheskyi zhurnal* 12 (1987): 77-80.

Barbasov, A. "Russkii voennyi deiatel' N. N. Obruchev." *Voenno-istoricheskyi zhurnal* 8 (1973): 100-105.

_____. "Novye fakty o planirovanii russko-turetskoi voiny 1877-1878 gg." *Voenno-istoricheskyi zhurnal* 2 (1976): 98-104.

Bonch-Bruevich, M. "M. I. Dragomirov. Osnovy vospitaniia i obrazovaniia voisk." *Voenno-istoricheskyi zhurnal* 3 (1973): 75-80.

D'iakov, V. A. "O razvitii russkoi voenno-istoricheskoi mysli v poslednei chetverti XIX veka." *Voenno-istoricheskyi zhurnal* 5 (1959): 60-72.

_____. "O vozniknovenii, sostave i deiatel'nosti russkogo voenno-istoricheskogo obshchenstva (1907-1917)." *Problemy istorii doshchestvennogo dvizhenie i sitoriografii,* pp. 275-88. Moscow, 1971.

Kavtaradze, A. G. "Iz istorii russkogo general'nogo shtaba." *Voenno-istoricheskyi zhurnal* 7 (1972): 87-92.

_____. "Voennye entsiklopedicheskie slovari i entsiklopedii dorevoliutsionnoi Rossii (1724-1915 gg.)." *Voenno-istoricheskyi zhurnal* 1 (1973): 105-9.

_____. "Iz istorii russkogo general'nogo shtaba (1909-iiul' 1914 gg.)." *Voenno-istoricheskyi zhurnal* 12 (1974): 80-86.

_____. "Iz istorii russkogo general'nogo shtaba (avgust 1914 goda-mai 1918 goda)." *Voenno-istoricheskyi zhurnal* 3 (1976): 103-9.

_____. "Voiskovoe upravlenie general'nogo shtaba russkoi armii." *Voenno-istoricheskyi zhurnal* 6 (1978): 77-82.

Khelme, R. "Voenno-teoreticheskie nasledstvie Zhomini." *Voenno-istoricheskyi zhurnal* 5 (1979): 88-89.

Kochetkov, A. "Russkii voennyi istorik M. I. Bogdanovich." *Voenno-istoricheskyi zhurnal* 8 (1968): 77-84.

Kozlov, S. "Voennaia nauka i voennye doktriny v pervoi mirovoi voine." *Voenno-istoricheskyi zhurnal* 11 (1964): 31–41.

Leonidov, L. "Voennyi istorik A. K. Puzyrevskii." *Voenno-istoricheskyi zhurnal* 1 (1969): 91–97.

______. "Iz istorii russkoi voenno-istoricheskoi mysli 60-x godov XIX veka." *Voenno-istoricheskyi zhurnal* 10 (1973): 95–100.

Rostunov, I. "Uroki russko-iaponskoi voiny 1904–1905 gg." *Voenno-istoricheskyi zhurnal* 2 (1984): 73–79.

Shaposhnikov, B. "V Varshavskom voennom okruge." *Voenno-istoricheskyi zhurnal* 1 (1967): 75–88.

Zhilin, P. "Diskussii o edinoi voennoi doktrine. (Iz istorii razvitiia voenno-teoreticheskoi mysli)." *Voenno-istoricheskyi zhurnal* 5 (1961): 61–74.

______. "Bol'shaia programma po usileniiu russkoi armii." *Voenno-istoricheskyi zhurnal* 7 (1974): 90–97.

Books, Non-Soviet

Alexinsky, G. *La Russie et la Guerre*. Paris: Librarie Armand Colin, 1915.

Berlin, I. *Russian Thinkers,* Harmondsworth: Pelican, 1978.

______. *Concepts and Categories: Philosophical Essays*. Oxford; Oxford University Press, 1980.

Bird, W. D. *Lectures on the Strategy of the Russo-Japanese War.* London, 1909.

Curtiss, J. S. *The Russian Army under Nicholas I, 1825–1855*. Durham: Duke University Press, 1965.

Craig, G. A. *The Politics of the Prussian Army, 1640–1945*. Oxford: Oxford University Press, 1964.

Donnet, G. *Histoire de la Guerre Russo-Japonaise*. 2 vols. Paris: Delagrave, 1904/5.

Duffy, C. *Russia's Military Way to the West*. London: Routledge and Kegan Paul, 1981.

Dziak, J. J. *Soviet Perceptions of Military Doctrine and Military Power: The Interaction of Theory and Practice*. New York: Crane, Russak, 1984.

Erickson, J. *The Russian Imperial/Soviet General Staff.* College Station, Tex.: Texas A & M University, 1984.

Fuller, W. C., Jr. *Civil-Military Conflict in Imperial Russia, 1881–1914*. Princeton: Princeton University Press, 1985.

Le Général Antoine-Henri Jomini (1779–1869). Contributions à sa biographie, 41. Lausanne: Bibliothèque Historique Vaudoise, 1969.

Goerlitz, W. *The History of the German General Staff, 1657–1945*. New York: Praeger, 1953.

Golovin, N. N. *The Russian Army in the World War.* New Haven: Yale University Press, 1931.

______. *The Russian Campaign of 1914*. Fort Leavenworth, Kans. 1933.

______. *Nauka o voine: o sotsiologicheskom izuchenii voiny.* Paris: Signal, 1938).

Historical Analysis of the Use of Mobile Forces by Russia and the USSR. Section I: "The Beginning: Imperial Russia and Soviet Mobile Warfare to 1920," by J. W. Kipp, pp. 15–115. Occasional Papers Series, 10. College Station, Tex.: Center for Strategic Technology, Texas A & M University, 1985.

Howard, M. *The Franco-Prussian War,* Methuen, 1979.

______. *Clausewitz*. Oxford: Oxford University Press, 1983.

Jomini and His "Summary of the Art of War," edited by J. D. Hittle. Harrisburg, Pa.: Stackpole, 1965.

Keep, J. L. H. *Soldiers of the Tsar: Army and Society in Russia, 1462–1874.* Oxford: Oxford University Press, 1985.

Kennan, G. F. *The Decline of Bismarck's European Order; Franco-Russian Relations, 1875–1890.* Princeton: Princeton University Press, 1979.

Kipp, J. W. *From Foresight to Forecasting: The Russian and Soviet Military Experience.* College Station, Tex.: Stratech Studies Series, Center for Strategic Technology, Texas A & M University, 1988.

Kroeber, A. L., and C. Kluckhohn. *Culture: A Critical Review of Concepts and Definitions.* Appendix A (Historical Notes on Ideological Aspects of the Concept of Culture in Germany and Russia) and Appendix B (The Use of the Term "Culture" in the Soviet Union) by Alfred G. Meyer. Papers of the Peabody Museum of American Archeology and Ethnology. Cambridge: Harvard University Press, 1952.

Leonard, R. A. *A Short Guide to Clausewitz–On War.* New York: Putnam's Sons, 1967.

Lieven, D. C. B. *Russian and the Origins of the First World War.* London: Macmillan, 1983.

Lincoln, W. B. *Nicholas I: Emperor and Autocrat of All the Russias,* Bloomington: Indiana University Press, 1978.

______. *In the Vanguard of Reform: Russia's Enlightened Bureaucrats, 1825–1861.* Dekalb: Northern Illinois University Press, 1982.

______. *In War's Dark Shadow: The Russians before the Great War.* New York: Dial, 1983.

Makers of Modern Strategy: From Machiavelli to the Nuclear Age, edited by P. Paret and M. Howard. Princeton: Princeton University Press, 1986.

Mayzel, M. *Generals and Revolutionaries. The Russian General Staff during the Revolution: A Study in the Transformation of Military Elite.* Osnabruck: Biblio Verlag, Studien zur Militargeschichte, Militarwissenschaft und Konflictforschung, 1979.

Miller, F. A *Dmitrii Miliutin and the Reform Era in Russia.* Nashville, Tenn.: Vanderbilt University Press, 1968.

Paret, P. *Clausewitz and the State: The Man, His Theories, and His Times.* Princeton: Princeton University Press, 1985.

Ponomareff, D. *Political Loyalty and Social Composition of a Military Elite: The Russian Officer Corps, 1861–1903.* Santa Monica, Calif.: Rand Corporation, 1977.

Porch, D. *The March to the Marne: The French Army, 1871–1914.* Cambridge: Cambridge University Press, 1981.

Pratt, E. A. *The Rise of Rail-Power in War and Conquest, 1833–1914.* London, 1915.

Presseissen, E. L. *Before Aggression: Europeans Prepare the Japanese Army.* Association for Asian Studies Monographs and Papers, 21. University of Arizona Press, 1965.

Roots of Strategy: A Collection of Military Classics, edited by T. R. Phillips. London: Bodley Head, 1943.

The Russo-Japanesse War. Prepared in the historical section of the German general staff. Translated by Karl von Donat. 4 vols. London: Hugh Rees, 1908.

Snyder, J. *The Ideology of the Offensive: Military Decision Making and the Disasters of 1914.* Ithaca: Cornell University Press, 1984.

Stone, N. *The Eastern Front, 1914–1917.* London: Hodder and Stoughton, 1975.

Weigley, R. F. *The American Way of War.* London: Collier, 1973.

White, C. E. *The Enlightened Soldier: Scharnhorst and the* Militarische Gesellschaft *in Berlin, 1801–1805.* New York: Praeger, 1989.
Whitton, F. E. *Moltke.* London: Constable, 1921.

Articles from Non-Soviet Periodicals

Arnold, J. C. "French Tactical Doctrine, 1870–1914." *Military Affairs* (Apr. 1978): 61–66.
Askew, W. C. "Russian Military Strength on the Eve of the Franco-Prussian War." *Slavonic and East European Review* 30 (1951/52): 185–205.
Bellamy, C. "Seventy Years On: Similarities between the Modern Soviet Army and Its Tsarist Predecessor." *RUSI Journal* (Sept. 1979): 29–38.
_____. "Heir of Genghis Khan: The Influence of the Tartar-Mongols on the Imperial Russian and Soviet Armies." *RUSI Journal* (Mar. 1983): 52–60.
_____. "Antecedents of the Modern Soviet Operational Manoeuver Group (OMG)." *RUSI Journal* (Sept. 1984): 50–59.
Brooks, E. W. "Reform in the Russian Army, 1856–1861" *Slavic Review* 43/1 (Spring 1984): 63–82.
_____. "The Improbable Connection: D. A. Miljutin and N. G. Cernysevskij, 1848–1862." *Jahrbucher fur Geschichte Osteuropas* 37 (1989): 21–43.
Collins, D. N. "The Franco-Russian Alliance and Russian Railways, 1891–1914." *Historical Journal* 16/4 (Dec. 1973): 747–88.
Dallas, D. I. "The Origin of Capital Staffs." *Journal of Modern History* 10/2 (June 1938): 161–79.
Deist, W. "Kaiser Wilhelm II in the Context of His Military and Naval Entourage." In *Kaiser Wilhelm II: New Interpretations,* edited by J. C. G. Rohl and N. Sombart, pp. 169–92. Cambridge: Cambridge University Press, 1982.
Erickson, J. "Koalitsionnaya Voina: Coalitional Warfare in Soviet Military Theory, Planning, and Performance." In *Coalition Warfare: An Uneasy Accord,* edited by K. Nielson and R. A. Prete, pp. 79–122. Waterloo, Ontario: Wilford Laurier University Press, 1983.
_____. "The General Staff: Theory and Practice from the Tsarist to the Soviet Regime." In *National Security Policy: The Decision-Making Process,* edited by A. L. Pfaltsgraff, Jr., and U. Ra'anan, pp. 15–85. Medford, Mass.: Archon, 1984.
Gilbert, F. "From Clausewitz to Delbruck and Hintze: Achievements and Failures of Military History." In *Strategy and the Social Sciences: Issues in Defence Policy,* edited by A. Perlmutter and J. Gooch, pp. 11–20. London: Frank Cass, 1981.
House, J. M. "The Decisive Attack: A New Look at French Infantry Tactics on the Eve of World War I." *Military Affairs* (Dec. 1976): 164–68.
Kanter, S. "Sacrificing National Defence to Class Interest: The French Military Service Law of 1872." *Military Affairs* (Jan. 1985): 5–8.
Kenez, P. "A Profile of the Pre-Revolutionary Officer Corps." *California Slavic Studies,* 7 (1973): 121–58.
Kipp, J. W. "Consequence of Defeat: Modernizing the Russian Navy, 1856–1890." *Jahrbucher Fur Geschichte Ost-europas* 20 (1972): 210–25.
Kovacs, A. F. "French Military Institutions before the Franco-Prussian War." *American Historical Review* 51 (1945): 217–35.

Lincoln, W. B., and J. W. Kipp. "Autocracy and Reform: Bureaucratic Absolutism and Political Modernization in Nineteenth-Century Russia." *Russian History* 6 (1979): 1-21.

Mayzel, M. "The Formation of the Russian General Staff, 1880-1917: A Social Study." *Cahiers du Monde russe et sovietique* (July-Dec. 1975): 297-321.

Menning, B. W. "The Deep Strike in Russian and Soviet Military History." *Journal of Soviet Military Studies* 1 (Apr. 1988): 9-28.

Messerschmidt, M. "German Staff Officer's Education since the Beginning of the 19th Century: Innovations and Traditions." *Militarhistorisk Tidskrift* (June 1983): 9-19.

Mitchell, A. " 'A Situation of Inferiority': French Military Reorganization after the Defeat of 1870." *American Historical Review* 86 (1981): 49-62.

Morrill, D. L. "Nicholas II and the Call for the First Hague Conference." *Journal of Modern History* 46 (June 1974): 296-313.

Perrins, M. "Russian Military Policy in the Far East and the 1905 Revolution in the Russian Army." *European Studies Review* 9 (1979): 331-49.

Reichel, D. "La position du Général Jomini en tant qu'expert militaire à la cour de Russie." *Actes du Symposium 1982,* Service historique, Travaux d'histoire militaire et de polémologie, vol. 1, pp. 59-75. Lausanne, 1982.

Showalter, D. "The Influence of Railroads on the Prussian Planning for the Seven Weeks War." *Military Affairs* (Apr. 1974): 62-67.

Stone, N. "Organizing an Economy for War: The Russian Shell Shortage, 1914-1917." In *War, Economy and the Military Mind,* edited by A. Wheatcroft and G. Best, pp. 108-19. London, 1976.

Turner, L. C. F. "The Russian Mobilization in 1914." *Journal of Contemporary History* 2 (1968).

UNPUBLISHED PAPERS AND DISSERTATIONS

Bilof, E. G. "The Imperial Russian General Staff and China in the Far East, 1880-1888: A Study of the Operations of the General Staff." Ph.D. diss., Syracuse University, 1974.

Bushnell, J. S. "Mutineers and Revolutionaries: Military Revolution in Russia, 1905-1907." Ph.D. diss., Indiana University, 1977.

Erickson, J. "On Investigating Soviet Military Institutions." Unpublished paper, Department of Defence Studies, University of Edinburgh, n.d.

Goldstein, E. R. "Military Aspects of Russian Industrialization: The Defense Industries, 1890-1917." Ph.D. diss., Case Western University, 1971.

Kipp, J. W. "Mass, Mobility, and the Red Army's Road to Operational Art, 1918-1936." Unpublished paper, Fort Leavenworth, Kans., Soviet Army Studies Office, U.S. Army Combined Arms Center, 1988.

______. "The Methodology of Foresight and Forecasting in Soviet Mililtary Affairs." Unpublished paper, Fort Leavenworth, Kans., Soviet Army Studies Office, U.S. Army Combined Arms Center, n.d.

Laney, F. M. "The Military Implementation of the Franco-Russian Alliance, 1890-1914." Ph.D. diss., University of Virginia, 1977.

Long, J. W. "The Economics of Franco-Russian Alliance, 1904-1906." Ph.D. diss., University of Wisconsin, 1968.

Malanchuk, G. V. "The Training of an Army: M. I. Dragomirov and the Imperial Russian Army, 1860–1905." M.A. thesis, Miami University, 1978.

Menning, B. W. "A. I. Chernyshev: A Russian Lycurgus." Unpublished paper, Fort Leavenworth, Kans., Soviet Army Studies Office, U.S. Army Combined Arms Center, 1984.

_____. "Army and Frontier in Russia." Unpublished paper, Fort Leavenworth, Kans., Soviet Army Studies Office, U.S. Army Combined Arms Center, n.d.

_____. "Bayonets before Bullets: The Imperial Russian Army, 1861–1914." Unpublished manuscript, 1989.

Pintner, W. "The Russian Army and Russian Society." Final report to the National Council for Soviet and East European Research, 1982.

Sloan, J. F. "Military History of Russia: A Preliminary Survey of Sources." Unpublished annotated bibliography, Garmisch, Germany, U.S. Army Institute for Advanced Russian and East European Studies, 1971.

Taylor, P. R. "The Trans-Siberian Railroad and the Russian Revolution of 1905." Ph.D. diss., University of Tennessee, 1969.

Von Wahlde, Peter. "Military Thought in Imperial Russia." Ph.D. diss., Indiana University, 1966.

Wilfong, W. T. "Rebuilding the Russian Army, 1905–1914: The Question of a Comprehensive Plan for National Defense." Ph.D. diss., Indiana University, 1977.

Index

Academic Conference, 5–6, 8, 12, 17, 34, 61, 63, 68; and admission procedures, 64; and annual "Report" of, 59; and the "applied method," 60; and creation of a course on Russian military art, 104; and debate over existence of Russian military art, 77; and examination of supplemental course theses, 66; and proctoring of supplementary course, 66; and reform of curriculum, 43; and reform of military history course, 65; and study abroad program, 44
Academic conferences, 5, 8; and increased responsibilities of, 59
Academic Council, 4–6, 8, 12, 17, 53; Miliutin's appointment to, 34; review of examination procedures, 35; under General Rostovtsev, 35; War Ministry criticism of low student enrollment, 34
Academic digest, 43
Academic dissertations, 44
Academic School, 100, 139
Academy (Nicholas Academy of the General Staff), xv, 161–62; as central military-scientific forum, 136; criticism of curriculum after Russo-Japanese War, 138; first review of curriculum, 34; and higher academic standards, 58; as a military-scientific forum, 5, 34, 136, 139, 152–54; and official regulations, 5, 8, 12, 59, 64; physical description of, 63; and referendum on military education, 136; and reform of, 35, 39, 54, 57; and relations with the French army, 120; and specialization under the Main Staff, 58
Adaridi, K., 154
Adrianopol, 80, 82–84
"Akademicheskii znak," 65
Alekseev, Admiral E., 131–32, 134
Alekseev, Major-General M. V., 135, 138, 145
Alexander I, 1
Alexander II, 43, 49, 55, 58, 80, 82, 83, 91; and approval of Miliutin's reform of troop organization, 54; assassination of, 92; and contradictory reform policies, 50; and end of Crimean War, 37; and establishment of the Ridiger Commission, 38; and support of the Imperial Russian Technological Society, 66

Alexander III, 92; and counter-reform of the Miliutin system, 92–93
Alexander Alexandrovich, Grand Duke, 83
Alexander the Great (of Macedonia), 13, 69, 71
American Civil War, 25, 65, 113
American Confederate Army, 113
Ancient Greeks, evolution of warfare, 10
Anichkov, V. M., 44, 50, 55; and editorial board of *Voennyi sbornik,* 50; as member of commission on Military Districts, 55; and military administration, 44
"Applicatory method," 65; as used by Dragomirov in tactics, 73. *See also* Applied School
Applied School, 59–61, 78, 154. *See also* "Applicatory Method"
Arab-Konak Pass, 84
Archives, and work of Maslovskii, 102. *See also* Department of General Archives of the General Staff, Military Educational Archives
Artamonov, N. D., 80
Artillery, theory of, 41
Artillery Academy, 50; and need for reform, 57
Artillery School, 5, 6
Astaf'ev, Colonel A. I., 39, 44, 67, 68, 76, 118; and bayonet/bullet analogy, 40; and criticism of unified theories, 41; and dual nature of military science, 40; and ideal model of higher military education, 41; and influence of V. G. Belinskii, 49; influence on Dragomirov, 73; and the Liberal School, 40; military administration as empirical science, 42; and military science as basis for administrative reform, 41–42, 45; and *On Contemporary Military Art,* 39; theoretical basis for future reform, 45; and unified theory of strategy, 40–41
Augustus II, 23
Austria (Austro-Hungarian Empire), 82; threat of military intervention during Crimean War, 37
Austro-Prussian War, 25, 65, 107; and Dragomirov's tactical study of, 68

Bagration, General, as portrayed by Tolstoy and critiqued by Dragomirov, 75
Baranov, Duke, 38
Barantsev, Major General, 38
Bariantinskii, Prince, 38, 54; and era of reform, 51; as member of the Ridiger Commission, 38; and opposition to Miliutin, 91
Baumgarten, A. K., 45, 51, 59
Bel'aev, General M. A., 152
Belinskii, V. G.: influence on Astaf'ev, 39; influence on military intelligentsia, 49
Berens, Colonel, 67, 68, 77
Berlin War Academy (Kriegsakademie), 1, 140; comparison of admission procedures with Nicholas Academy of the General Staff, 64
Bielostok, 122
Bliokh, I. S., 114, 115, 116; and economic (statistical) analysis of modern warfare, 114; and *General Conclusions from "Future War in Its Technological, Economic and Political Aspects,"* 113; and influence on Gulevich, 116; and influence on Mikhnevich, 143; on officer training, 114
Blume, General, 140
Bobrovskii, P. O., 106
Bobr River, 152
Bogatyrs, 23. *See also* Genius; Military art, theory of
Bogdanovich, 17, 25, 34, 45, 61, 67–68; and "About the newly perfected hand fire-arms," 42; on Clausewitz's work *On War,* 23; criticized by Astaf'ev, 39; and development of military historical archives, 67; and *Essays on Strategy: Rules of Conducting War,* 23; and *History of the Patriot War of 1812,* 43; and the Imperial Russian Geographical Society, 34; influence by Clausewitz, 23; influence of Crimean War on, 42; influence of Medem on, 22; influence

on G. A. Leer, 68; and "On the Influence of the Technical Aspects of Military Affairs on Tactics," 42; and "On the Study of Military History," 42; and participation on *Military Encyclopedic Lexicon*, 22–23; on theory of military art, 23–24, 42
Bol'shaia Programma, 152. *See also* Great Military Program
Bonch-Bruevich, M. D., 154, 161
Borets, A., 154
Bug River, 122, 152; planned retreat of Russian army to, 80
Bulgaria, 80, 82
Bulow, A. von, 11

Central Asia, 143; and Maksheev's courses on, 68
Charles II, 23
Chebyshev, V. L., 57
Chernishevskii, N. G., 49–50
Chernyshev, General, 5
Chief of the General Staff: and Bariantinskii's reform proposal, 51; under Obruchev, 94; under General Palitsyn, 134. *See also* General Staff, Main Administration of the General Staff
Chief of the Main Staff, 4, 135; proposed increase in status, 93; under Prince Bariantinskii, 91; under General Chernyshev, 5; under Obruchev, 92. *See also* Main Staff
China, 68, 135
Clausewitz, C. von, xiv, 11, 18, 24, 25, 26, 69, 70, 74, 142; and the "applied method," 60; criticisms, of Jomini's work, 13; influence on strategic theory of G. A. Leer, 68; on strategy as art/science, 71
Coalitional warfare, 116
Committee on Structure and Training of Troops, 75, 97
Conference of Constantinople, 82
Congress of Berlin, 84
Congress of Vienna, xiii, 1, 35, 37
Constantinople, 80, 82, 83, 84
Contemporary war, 141, 145; and abolition of, 151; and "attritional" nature of, 148; and concept of "orientirovaniia," 150; Council for State Defense, 134–35; as evaluated by Gerua, 140; and formulation of Field Regulations (1912), 153; and significance of railroads, 147
Council of State Defense, 134–35
"Coup d'oeil," xiv, 97; as conceived by Dragomirov, 75. *See also* Glazomer
Crimean War, 10, 33, 35, 39, 41, 42, 43, 45, 50, 54, 58, 64, 65, 70, 74, 79, 80, 96; as analyzed by Leer and Dragomirov, 44; critico-historical analysis and selection of strategic goals, 100; as example of cultural renewal by warfare, 69; and importance of modernized weaponry, 56; and influence on the teaching of military history, 67; and reform of officer education, 57; strategic planning during, 78; technology and evolution of military art, 37, 39
Critico-historical method, 69–72; and theories of Gerua, 140
Curriculum (of the Nicholas Academy of the General Staff), 5; auxiliary subjects, 63; differentiation of theoretical and practical courses, 44; and first-year courses, 64; instruction in foreign languages, 65; practical courses, 4–5, 45; preparatory courses, 4–5; secondary courses, 65; supplementary courses, 65–66; and Zeddeler's influence on, 8

Danilov, General, 152
Dannenburg, P. A., 38, 54
Danube River, 80, 82
Danubian Army, 82
Department of General Archives of the General Staff, 67
Department of Higher Military-Educational Institutions, 44
District councils, 55, 56
Dobroliubov, N. A., 49, 50
Dobrorolskii, Colonel, 135, 145

Doctrine, xv, 161–162, 164: first systematic attempt at, 91; of the Plevna School, 96. *See also* Military art, theory of
Dolgorukov, V. A., 50
Domanevskii, V., 153–54; and "On the New Field Service Regulations," 153; and "Unity of Military Doctrine," 153; and "Unity of Military Doctrine and the Field Regulations," 153
Dragomirov, M. I., 101, 114; as chief of staff for the Kiev Military District, 76; as commander of the Kiev Military District, 99; and command of the 14th Infantry Division, 83; and concept of the nature of warfare, 74; criticism of Kuropatkin, 137; and criticism of I. S. Bliokh's work, 115; and debate over tactical doctrine, 96; on the deficiencies of higher military education, 94; description of early career, 73; as Director of the Academy, 93–94; and dynamic new era for the Academy, 94; as evaluated by Gerua, 140; and "extreme" tactical exercises, 97–98; and Faculty of Tactics, 44; and influence on Bliokh, 114; and influence on Neznamov, 144, 151; and influence on post-1905 theory, 148; and influence over formulation of Field Regulations (1881), 97; as member of commission on review of Miliutin system, 92; and opposing theories, 76; and post-Russo-Japanese War, 141; and reform of military history, 67; retirement from directorship of Academy, 104; and review of Tolstoy's *War and Peace,* 75; and the significance of the battle for Plevna, 84; and sponsorship of Puzyrevskii, 103; and study-abroad program, 44, 73; and support for Military District system, 93; and tactical style during grand maneuvers, 98; and the teaching of tactics, 75; and theory of military art, 72; and the "Traversant maneouver à la Souvoroff," 97
Duma, 152

Eastern problem, 80. *See also* Russo-Turkish War of 1877-1878
Ecole Superieure Militaire, 1, 152; and Obruchev's study at, 60
"Edinonachale," 109, 148, 149, 150, 162–63. *See also* Military art, theory of
Emir Pasha, 35
Encyclopedia of Military and Naval Science, 94, 99
Engalichev, General: as Director of the Academy, 154; and reform of central administration, 134
Engineering Academy, 51, 57, 68
Engineering schools, 5
Entente Powers (Crimean War), 35–37
Ermolov, General, 136
Examination Commission (of the Nicholas Academy), 64

Fadeev, General, 91
Feldman, Major-General, 120–21
Field Regulations (1879), 96
Field Regulations (1881), 96–97, 119
Field Regulations (1912), 153, 161
Filipopol, 84
Foresight. *See* Military art, theory of
Fortresses, strategic significance of, 147
Franco-Austrian War, 50, 79
Franco-Prussian War, 25, 65, 76, 77, 107; and influence on Leer's theories, 76; and shift of European military-political power, 78
Franco-Russian Military Convention, 122
Frederick the Great, 11, 13, 18, 74
French Army, relations with Russian High Command, 120
Frunze, M. V., 161
Future war, 114. *See also* Bliokh, I. S.

Gaillard, Colonel (French Military Attaché), 120
Galicia, 121–22
Gazenkampf, General, 94
Gechevich, Major General, 38
Geiden, General, 55
Geisman, P. A., 105, 112, 115; and adoption of Puzyrevskii's comparative

cultural method, 106; and *Attempt at a Study of the Tactics of Mass Armies,* 106; and concept of military art, 106; and concept of "orientirovaniia," 150; and importance of education, 109; on the importance of military history, 105; and reappraisal of military education, 136; and *Short Course in the History of Military Art in the Middle and Modern Centuries,* 105
General plan of state defense, 137, 145; and significance of railroads, 147
General Staff: Astaf'ev on the new responsibilities of, 42; consolidation of functions, 54; Department of, 54; importance of, 4; importance of education in new strategic era, 109; and importance of scientific training, 113; as link between commander and army, 42; new emphasis on training, 43; and new professional responsibilities, 63; as practitioners of "orientirovaniia," 150; professional ethic, 4; training of, 3
Genius, xiii, 12, 14, 16, 23, 25, 69, 99–103, 108–10, 140, 142; as defined by Leer and Geisman, 105; Napoleon as symbol of, 64. *See also* Military art, theory of
Gershal'man, F., 122
Gerua, Colonel A. V., 140–42
Glazomer, 70, 106, 146, 149. *See also* Military art, theory of
Glinka, Major General, 38
Glinoetskii, N. P., 44
Golitsyn, N. S., 106; and influence on Puzyrevskii, 103; as professor of strategy, 17; reform of course on strategy, 16; and unified strategic theory, 17, 22
Golovin, General N. N., 162–63
Goltz, C. von der, 111
Goremikin, F., 17, 22, 24, 35; acceptance of strategic duality, 23; and *Guide to the Study of Tactics,* 26; influence on Dragomirov, 73; on military science, 25; as professor of tactics, 25; on theory of tactics, 26
Great Military Program, 146
Grigor'ev, V., 154
Grippenberg, General, 132
Gudim-Levkevich, Colonel P. M., 66, 77, 98; and *Historical Development of Russia's Armed Forces to 1708,* 77
Gulevich, A., 115, 117; influence on Mikhnevich, 143; and *War and National Economy,* 115
Gurko, General V. I. 83–84, 122, 135; as commander of Vilna Military District, 98; and proposals for reform of Main Staff, 123; and strategic role of the Russo-Polish Army, 122

Hannibal, 115
Harbin, 132–33
Hegelian philosophy, 69, 109
Heihachiro, Admiral Togo, 131
Herzen, A., 49
Historical School, 11
History of famous campaigns, 64, 102. *See also* Military history
History of military art, 8, 64, 76, 102; Astaf'ev's high opinion of, 39; and career of A. K. Puzyrevskii, 101; and P. A. Geisman as professor of, 105; as synthesis of historical processes, 102; under G. A. Leer, 62, 77, 94. *See also* Military art, theory of
Hungarian Rebellion of 1849, 65

Ianushkevich, General, 154
Imperial Russian Geographical Society, 22, 33
Imperial Russian Technical Society, 57, 66
Inspectorate Department, 38, 41, 54
Institute of Engineering and Communications, 6, 12
Intelligence, in contemporary war, 149
Isakov, Major-General, 58, 60
Izmest'ev, P. N., 161

Janin, Colonel, 152
Japanese Army (1st), 131–32
Jomini, Antoine-Henri, xiv, 1, 3, 4, 5, 11, 14, 15, 16, 19, 24, 25, 42, 99,

115; and Academy as scientific forum, 26; and geometry of military art, 11; and the Historical School, 11; and observations on the applied method, 60; and role of politics in strategy, 18; and Special Commission, 5; and strategic theory, 11; and *Summary of the Art of War,* 13; and *Treatise upon Grand Military Operations,* 12
Journal of the IRVIO, 136

Kalakutskii, N. V., 57
Kant, I., 69, 111
Kartsov, A. P., 34, 35, 84; and acquaintance with Chernyshevskii, 50; as member of the Ridiger Commission, 38; and Special Commission, 43–44
Kaulbars, General, 132
Kazanskii, Colonel, 66
Khanykov, Ia. V., 34
Khatkov, General, 5
Khoinkioi Pass, 83
Koliubakhin, B., 146
Ko'shmidt, Colonel V. B., 137
Kotseba, General, 92
Kridner, General, 83
Kriegsakademie. *See* Berlin War Academy
Krivtsov, E., 154
Kurdiumov, Major General, 38
Kuroki, General, 132
Kuropatkin, General A. N., 96, 117, 123; appointment as War Minister, 116; and command during Russo-Japanese War, 131–134; criticism of, 137; and underestimation of Japanese army, 137
Kutuzov, General, 43, 75

Lavrov, P. L., 50
Lebedev, P. S., 34, 43–44
Le Comte, A., 142
Lee, General, 113
Leer, G. A., 44, 68–72, 77, 98–100, 104–6, 107, 110, 111, 114, 115, 118, 119, 139–40; and Academic School, 102; as criticized by Puzyrevskii, 101; and the critico-historical method, 61; as general editor of the *Review of Russia's Wars from Peter the Great to Our times,* 94; and influence on post-1905 theory, 141, 148; and influence on Puzyrevskii, 103; and philosophical nature of war, 69; and re-evaluation of Field Regulations (1879 and 1881), 96; and reform of military history, 67; and reform of tactical exercises, 94; retirement from directorship of Academy, 117. Works: *Attempt at a Critico-Historical Analysis of Laws of the Art of Conducting War,* 108; *Attempt at a Critico-Historical Approach Towards a Positive Strategy,* 68, 76, 142; *Comparison of a Few Evolutions in Strategic Art,* 105; *Complex Operations,* 104; *Strategiia,* 99, 116
Le Flo, General (French Ambassador to Russia), 120
Leont'ev, General, 59, 66, 93
Lewal, General (French military theoretician), 150
Liao-tung Mountains, 132
Lia-yang, 131–32
Liberal School, 20, 24–26, 33
"Lichnost." *See* Genius; Military art, theory of
Linievich, General, 132–33
Liprandi, I. P., 43
Lloyd, H., 11, 15, 70, 71, 72
Lobko, P. L., 94, 123
London Protocol, 82
Loshkarev, Major-General, 38
Louis XIV, 109
Lovcha, battle of, 83

Mahon, Captain (French military observer), 122
Main Administration of the General Staff, 54; Department of Structure and Duty of Troops, 153; and intelligence duties, 149; Military-Historical Department, 134; Operations Department, 145
Main Artillery Administration, 57
Main Chief of Military-Educational Institutions, 35
Main Staff, 54, 80, 120; and formulation of Plan of Concentration No. 1, 121; Mobilization Section, 152; and strategic planning, 123

Maksheev, F., 43, 45, 67, 96
Maksimovich, Lieutenant-General, 38
Manchuria, 131, 141
Manchurian Army, 131–37
Martynov, E. I., 109–15, 134, 137; and criticism of Leer-Dragomirov theoretical heritage, 139; on the leadership style of Napoleon, 108; Petrov's criticism of, 112; and redefinition of military art, 107; and *The Sorry Experience of the Russo-Japanese War,* 137; *Strategy in the Epoch of Napoleon and in Our Time,* 106–7
Marxism-Leninism, 162
Maslov, I. E., 112, 146
Maslovskii, D. F., 94; and debate with Leer, 101–2, 104; and definition of military art, 102; and distinct Russian military art, 103; influence on Martynov, 107; influence on Mikhnevich, 118
Meckel, Colonel (German military theoretician), 132
Medem, Baron von, 6, 8, 10–12, 16, 17, 18, 19, 20, 22, 23, 25, 26, 40, 41, 102; and influence on Astaf'ev, 39; influence on Yazykov's work, 15; on Jomini's theory, 11; retirement from the Academy, 17; and *Review of the Know Laws and Systems of Strategy,* 10, 13, 17; and *Tactics,* 11
Men'kov, General P. K., 50, 60, 61
Merkhelevich, Lieutenant-General, 38
Mikhail Nikolaevich, Grand Duke, 57–58
Mikhailovskii-Danilevskii, A. I., 34
Mikhail Pavlovich, Grand Duke, 6
Mikhnevich, N. P., 97, 107, 142–44, 153, 161–62; and appointment as Director of the Academy, 117; and *The Basis of Russian Military Art,* 117; as commander of 5th Army Corps, 141; and debates over strategic planning, 146; and evaluation of military education, 136, 139; and evolution of military art, 143–44; and *Influence of new technological innovations on the tactics of troops,* 117–18; and nationalism in military science, 142; and *Strategiia,* 117; and theory of military art, 117; and unified strategic theory, 119
Militargesellschaft, xiv
Military-Academic Committee of the General Staff, 67
Military administration, 4, 44, 62, 64, 65; Astaf'ev's course, 42; course on, 94; and influence of I. S. Bliokh on, 115; under Rediger, 116; as taught by Lebedev, 34; theory of, 41
Military art, theory of, xiii–xvi, 3, 8, 11, 15–16, 18, 23, 148, 163; Astaf'ev and, 39–40; as a basis for administrative reform, 39; and careerism, 137; and "clerical" style of leadership, 137–38, 150; components analysed by Leer, 71; and concept of "democratization," 108–9, 114, 138, 148; and concept of disinformation, 149; and concept of "economy of force," 143, 146–47; and concept of "edinonachale," 148, 149; and the concept of "glazomer," 106; and concept of "lichnost," 23; and concept of "orientirovaniia," 106, 138; and concept of "predvidenie" (foresight), 149; and concept of secrecy, 149; as defined by Puzyrevskii, 101; materialization, numeralization, and democratization, 106–8, 164; and nationalist interpretations, 77; and prediction, 162–63; and theory of "borrowing," 103–4; and theory of "national rules," 106
Military Council, 134
Military diagnostics, 70
Military Districts, 19, 53, 54, 56, 62, 64, 84, 97, 120, 123, 141; attempted counter-reform of, 91–93; and Chief of, 93; as component in theory of military administration, 42; and conduct of annual grand maneuvers, 98; formation of, 54–56; as military scientific forums, 135; and Miliutin's reform proposal under Ridiger Commission, 38; and need for better officer education, 61; problems in administration during Russo-Turkish War, 96; strategic planning and mobilization, 78–80; tactical instruction at the St. Petersburg, 65
Military District staffs, 149

Military doctrine, 4, 91
Military Educational Archives, 136
Military-Educational Institutions, 54
Military Encyclopedia Lexicon, 16–17, 20
Military engineering, 41
Military fortification, 4
Military geography and statistics, course on, 4, 12; components of military policy, 16; Miliutin's reform of course, 21; under Miliutin, 20; under P. A. Yazykov, 6
Military geography, theory of, 12–16, 41; new emphasis in General Staff education, 42; Review of Empire, 22; and selection of strategic goals, 19, 100
Military-Historical Commissions, 135, 138
Military-Historical Department, 135, 136
Military history, xv, 3, 4, 6, 8, 41, 45, 139, 152
Military pedagogy, 68
Military policy, 13. *See also* Military geography and statistics, course on
Military science, 3, 5, 8, 25, 99, 136, 140, 141, 146; Astafev's definition of, 40–41; Leer's concept of, 69; and Martynov's concept of, 109; and Mikhnevich's concept of, 142; military psychology as new component of, 71; re-evaluation after Russo-Japanese War, 139; and relationship to military art, 12; Yazykov and the speculative nature of science, 15
Military-Scientific Commission, 78
Military-Scientific Committee of the General Staff, 67
Military Statistics, 41, 64; and additional lecture material, 67; course on, 62; definition of, 20–22; Obruchev and the applied method, 60; reformed course of Maksheev, 67; under Maksheev and Obruchev, 44–45. *See also* Military geography and statistics, course on
Military-Topographical Section of the General Staff, 54
Military topography, 41
Miliutin, D. A., 17, 20–22, 24, 25, 34, 35, 41, 42, 44, 45, 94, 118; acceptance of strategic duality, 23; and acquaintance with Chernyshevskii, 50; and annual training maneuvers, 97; appointment as court historian, 34; appointment to the Ridiger Commission, 38; as Chief of Staff for the Caucasian Army, 38; and covert diplomacy during Russo-Turkish War, 83; criticism of Yazykov, 22; and failure of counter-reform movement, 93; and *First Attempt at Military Statistics*, 21, 25; and *History of the War in 1799 between Russia and France in the Reign of Paul I*, 34; on improvement of academic standards, 35; on Medem's contributions to theory, 11; as member of the Imperial Russian Geographical Society, 34; as member of the Liberal School, 20; and military reforms, 51–61; and new priorities of strategic planning, 120; and Obruchev's war plan for the Russo-Turkish War, 80; and opponents to reform, 91; and rationalization of strategic planning, 78; reform proposal under Ridiger Commission, 38; and resignation as War Minister, 92; and "statistical" method, 21; and strategic defensive in strategic planning, 143; strategic planning during Crimean War, 78; on General Suvorov, 20; theoretical basis for future reform, 45
Miliutin, N., 34
Ministry of Interior, 34
Ministry of the Navy, 57
Ministry of Trade and Manufacturing (Department of Mining), 57
Mobilization, 80; and comparative disadvantage with other European powers, 79; lack of instruction in supplementary courses, 99; mobilization plan (Plan No. 19), 152. *See also* Military art, theory of; Military geography, theory of
Mobilization Section of the Main Staff. *See* Main Staff
Moltke, General H. von (the Elder), 76, 107–8
Montesquieu, C.-L., 115
Montholon, General, 15

Morava River, 121
Moulin, L. (French military attaché), 121, 122
Mukhden, 132
Murav'ev, General, 1
Myshlaevskii, A. Z., 117, 136

"Nakhodchivost," 69. *See also* Military art, theory of
Napoleon I, xiii, 11, 13, 23, 24, 63, 64, 69, 70, 76, 78, 99, 100, 107–8, 110, 115
Napoleon III, 50
Napoleonic Wars, 40, 105
Narev River, 80, 122, 152
National School, 107; after the Russo-Japanese War, 139; and criticism of Neznamov's call for a unified military doctrine, 145
Nation-in-arms, concept of, 146, 148
Neelov, Captain N. D., 24–25, 118; and *A Study on the Composition of Contemporary Strategy,* 24
Neidgardt, General-Adjutant, 5
Neiman River, 152
Neznamov, A. A., 146, 154, 161–62; and concept of contemporary war, 145, 151; and concept of foresight, 149; and concept of orientirovaniia, 150; and *Contemporary War. The Conduct of Field Armies,* 144; and democratization of military art, 138; and importance of reconnaisance, 149; influence of Dragomirov on, 151; and theories of contemporary military art, 144; and a unified military doctrine, 144, 145
Nicholas I, 1, 3, 5, 13, 33, 35, 37, 50
Nicholas II, 116, 134, 151, 154
Nicholas Academy of the General Staff. *See* Academy
Nikolai Nikolaevich, Grand Duke, 120, 134–35, 151, 162
Nikopol, 83
Nogi, General, 132
Novitskii, Colonel V. F., 139, 146, 161

Obruchev, General N. N., 55, 83, 84, 93, 117, 123; and acquaintance with Chernyshevskii, 50; and antagonism with Grand Duke Nikolai Nikolaevich, 120; and the applied method, 44, 60, 65; as chief of the Main Staff under Vannovskii, 92; as chief of the Military-Scientific Committee of the General Staff, 67; and confidence in professionalism of Academy graduates, 65; and "Considerations on the defense of Russia," 78; and course on Military Statistics, 94; on the defense of Warsaw, 113; and editorial board of *Voennyi sbornik,* 50; and final attempt to reform War Ministry, 123; and ideological break with revolutionary movement, 51; and introduction of staff rides, 78; as member of commission on review of Miliutin system, 92; on need to preserve Miliutin system, 92; and "Our Political Position at the Present Time," 82; as professor of Military Statistics, 45; relationship with the French Army, 121; and relations with French high command, 122; resignation of, 116; and "Strategic considerations of the war plan," 80; and strategic planning, 78–80, 82, 120–24, 143; study abroad program, 44; and support for Military District system, 93
Oder River, 121
Official Russian Military-Historical Society (IRVIO), 135, 136
Ogarev, N. P., 49
Ogorodnikov, F., 154
Okunev, Colonel N. A., 5, 12
Operational planning, 149, 151
"Orientirovaniia," 150
Osman Pasha, 83, 84
Ottoman Empire, 1, 35, 80, 82
Oyama, Marshal, 132

Palitsyn, General F. F., 134–36, 151
Patriotic War of 1812, 12, 43
Persia, 68
Peter the Great, 23, 63, 103, 106
Petrov, A. N., 109; criticism of Martynov, 113; and implications of technology on leadership, 113; influence on

Mikhnevich, 118; and *Tasks of Contemporary Strategy: As regards the work of Captain Martynov, Strategy in the Era of Napoleon I and in Our Time,* 109, 111, 112; and theory of military art, 110–12, 151
Piktorov, General, 55
Pin-yang, 131
Plan of concentration (No. 1), 121
Plevna, 83, 84, 118, 132; battles at, 83; and significance for tactical theory, 96
Plevna School, the, 96
Polish Crisis of 1863, 79
Polish Salient, 122
Polivanov, General, 55
Port Arthur, 131–32
Proudhon, P. J., 111, 115
Prussian Silesia, 121
Prusso-Austrian War of 1866, 77
Puzyrevskii, General A. K., 77, 98, 106, 111, 114, 117, 121; appointment as Chief of Staff for the Warsaw Military District, 105; and criticism of Bliokh's work, 115; and *Development of Continuous Regular Armies and the Composition of Military Art in the Century of Louis XIV and Peter the Great,* 103; and evaluation of Field Regulations, 96; and *History of Military Art in the Middle Ages,* 101; influence on Bliokh, 114; influence on Geisman, 105; influence on Martynov, 107; influence on Mikhnevich, 118; and "Memoirs of an Officer of the General Staff on the War of 1877–78 in European Turkey," 101; as observer of French grand maneuvers, 120; opposition to Leer, 101, 104; and relations with French high command, 122; and reliance on Stankevich's work, 101; and strategic calculations, 122; and theory of military art, 94, 101–3; and a uniquely Russian military art, 102–3; and "War Games and a Guide to their Conduct," 66

Queen Victoria, 83

Radetskii, General, 84
Railroads. *See* Strategic Planning; Strategy, theory of
Rediger, A. F., 94, 96, 116; and *Build-up and Formation of the Armed Forces,* 116; and resignation as War Minister, 152; as War Minister, 151
Reform (Higher Military-Educational Institutions), 57, 60
Regulations of the Field Administration of Armies in War Time, 96, 123
Reinsurance Treaty (between Germany and Russia), 122
Rennenkampf, General, 34
Review of Russia's Wars from Peter the Great to Our Times, 94
Ridiger, General, 37–38, 41
Ridiger Commission (Commission for the Improvement of the Military Element), 38, 43, 49, 50, 53, 54, 56
Romania, 80
Rostovtsev, General, 35; reform of Academy's examination system, 43; and study abroad program, 44
Routine, xv
Rudtorfer, Lieutenant Colonel, 12, 14–15
Rumiantsev, General, 23
Rushchuk, 80
Russian Civil War, 162
Russian-French military relations, 121
Russian military art, 103, 106. *See also* Military art, theory of; Puzyrevskii, General A. K.
Russian National School, 107
Russkyi invalide, 66, 153
Russo-Japanese War of 1904–1905, 94, 98, 107, 116, 135, 137, 146, 150, 151; and crisis of leadership, 148; Neznamov's participation in, 144
Russo-Polish Army, 122
Russo-Polish War of 1830–1831, 5, 65, 77
Russo-Turkish War of 1877–1878, 64, 65, 76, 77, 92, 99, 101, 116, 118, 119, 120, 135; analysis of, 96; and Petrov's analysis of, 110–11
Ruzskii, General, 153

Sakharov, General, 117, 134
Scharnhorst, General J. H. von, xiv, 60
School for Column Leaders, 1
Schubert, Major-General, 5
Science, 11, 16, 20, 163. *See also* Military Science
Scientific positivism, 68, 142
Semenov-Tian-Shanskii, P. P., 34
Sevastopol (Crimean War), 35, 37
Seven Years War, 10
Sha-ho River, 132
Shcherbatchev, General, 152, 154
Shenshin, Captain, 77
Shipka Pass, 82, 83, 84
Shipka-Sheinovo, 93, 118–19
Shtang, General, 55
Shumla, Fortress of, 80
Silistria, Fortress of, 80
Sistovo, Fortress of, 80, 82, 83
Skalon, General, 136
Skobelev, General M. D., 84, 96, 116, 118, 119, 145; and Kuropatkin's style of command during Russo-Japanese War, 132; as member of commission review of Miliutin system, 92
Skugarevskii, A. P., 102
Small Arms Commission, 57
Social-Darwinism, 69
Society of the Enthusiasts of Military Knowledge, 135
South Western Army (Russian), 122
Special Commission, 5, 6, 26; and adaptation of the Applied School, 59; on hiring policy for professors, 43; and opposition to Miliutin's reforms, 92; and Miliutin's reform of the military, 51; on post-Miliutin reform of education, 94; and re-evaluation of Field Regulations (1879 and 1881), 96; on reform of higher education, 60; and review of Military District system, 93
Special Committee, 54, 58
Special Conference, 120
Special Meeting, 63, 67, 68, 77
St. Cyr, 1, 60, 64, 73
Staff rides, 65, 78
Stankevich, Colonel, 66, 67, 68, 71, 101; and course on military art, 77; death of, 94; and *History of Military Art*, 77; and post-Miliutin reform of education, 94; and reform of military history, 67
Stavka, 148
Stefan, G. F., 12, 13; appointment as Director of Academy, 35; member of the Imperial Russian Geographical Society, 34; and post-Crimean War reform of curriculum, 43; as Vice-Director of Academy, 34
Strategic defensive, 143, 147
Strategic exhaustion, 143
Strategic offensive, 146
Strategic planning, xv, 19, 22, 25, 77, 120, 122, 123, 137, 146, 147, 151; and close relations between Russian and French high command, 122; and concept of force concentration, 79; during the Crimean War, 35; and evolution of "operational planning," 148; and influence of Von Der Goltz on Petrov, 111; Miliutin's theory of, 22; reform of, 124; and significance of railroads, 79, 147
Strategy, Military History and Military Literature, Faculty of, 4, 6, 8, 10, 24, 67, 139, 144
Strategy, theory of, xiii–xiv, 3, 4, 8, 10, 17, 41, 64, 65, 76, 141; analytical role of history, 15; as art/science, 10, 14, 23, 71; Astaf'ev's concept of, 40; and Berens'; lectures on, 67; as component of behavioral sciences, 69; components of, 11; Goremykin's contribution to theory of, 26; and instruction on the conduct of a "strategic review," 100; under G. A. Leer, 44, 62; and Leer's redefinition of, 105; and Medem's theory, 11; and role of chance, 71; and use of railroad, 25, 79. *See also* Military art, theory of
Sukhomlinov, General, 146, 152
Sukhotin, Lieutenant-General, 77, 98; and course on the history of Military Art, 94; and debate with Leer, 101, 104
Sukhozanet, General-Adjutant I. O., 5, 6,

17, 34, 45; appointed as Director of the Academy, 6; and conflict with Zeddeler, 8, 10; dismissal as Director, 35
Sukhozanet, N. O., War Minister, 50
Sukhtelen, Count, 5
Suleiman Pasha, 83, 84
Supplementary course, 65, 66; and determination of the "lines of operation," 99; influence of Leer over the curriculum, 99; publication of the *Synopsis of Strategy*, 99; and reliance on Napoleonic legacy, 100
Supreme Military Council, 93
Suvorov, General A. V., 10, 20, 23, 34, 40, 72, 75, 97, 116, 145
Svechin, A. A., 98, 161
Synopsis of Strategy, 99–100

Tactical exercises 97, 98
Tactical map exercises, 65
Tactics, course on, 4, 6, 8, 12, 24–25, 34, 44–45, 65, 68, 75, 139
Tactics, theory of, 41, 64, 65, 96; Astaf'ev's concept of, 40; Medem's definition of, 11; under M.I. Dragomirov, 73
Technical Committee of the General Staff, 120
Technology, 23, 39, 108, 118–119, 141; and civil-military scientific societies, 66; communications and the crisis of leadership in post-1905 theory, 148; and criticism of Russian equipment during Russo-Japanese War, 138; evaluation during grand maneuvers, 98; influence on theory of Military Geography, 15; and Petrov's criticism of Martynov's theories, 112; significance of railroads to strategic planning, 147; and subjective/objective balance of military science, 40. *See also* Military art, theory of
Territorial administration. *See* Military Districts
Theater of Military Operations, 71, 99, 105, 108, 123
Theater of War, 3, 13, 19, 20, 22, 25, 54, 55, 66, 78–79, 137, 145–47, 149
Theoretical balance, 70
Tirnovo, fortress of, 80, 82, 83
Tolstoy, Count L., 78, 112; criticism of Bogdanovich's work, 43; on the mystery of military art, 69, and *War and Peace,* 43, 69, 74, 75, 78, 112
Triandafillov, V. K., 161
Trionov Pass, 84
Triple Alliance, 145
Tsushima Straits, 133
Tukhachevskii, M. N., 161
Tur-en-chen, battle of, 132, 146
Turkey (Ottoman Empire), 82

Unified military doctrine, 99, 139, 154; and concept of "orientirovaniia," 151; and controversy over formulation of Field Regulations (1912), 153; and Domanevskii's call for a tactical language, 153; and Neznamov's call for, 145
Universal male conscription, 114, 116
University of St. Petersburg, 65
Ustrialov, General, 55

Vannovksii, General, 92–93; and approval of new course on Russian Military Art as War Minister, 104; as member of commission on review of Miliutin system, 92; resignation of, 116, 123
Varna, 80
Viemarn, Colonel, 6, 25
Vilna Military District, 98, 135
Visla River, 79–80, 122
Vladivostok, 131
Voennyi sbornik, 50, 62, 68, 69, 76, 104, 109, 112, 115, 117
Voennyi zhurnal, 42
Voenspetsi, 161
"Vsepoddaneishyi doklad," 53–57

War games, 66, 163
War Ministry, 51, 78, 80, 91–92, 93–94, 104; academic reform during Crimean War, 35; and Bariantinskii's reform proposal, 51; as central component in theory of military administration, 42; under General Chernyshev, 5; criticism

of Academy, 34; and Obruchev's proposals for reform, 123; post-1909 reform of, 151, 152; and "Provisional Storehouse Department of the Manchurian Army," 135; and rearmament program, 57; and reform of administrative functions, 55; reform of central administrative functions, 53 and report on reform of higher education, 61; and support of Leer's critico-historical method, 61; and temporary abandonment of Obruchev's plan, 83; and *Voennyi sbornik,* 50

War plan. *See* Strategic planning

Warsaw Military District, 117; as a military scientific center, 114; and Puzyrevskii's appointment as Chief of Staff, 105; and relations with French high command, 122; and strategic analysis by the Statistical Section, 122; and strategic planning, 122

Waterloo, xiii

White Russian Army, 162

Wrangel, General, 162

Wellesley, Colonel (British Military Attaché), 83

Ya-lou River, 131–32, 146

Yakimovich, General, 55

Yazykov, Major P. A., 6, 8, 12, 23, 24, 34, 41, 118; and *Attempt at a Military Geography,* 13; and *Attempt at a Theory of Strategy,* 17; on education of General Staff officers, 16; and inherent offensive or defensive nature of a nation's armed forces, 18; influence on Leer, 70; and Jomini's theoretical principles, 13; and Positivist orientation, 19; praise of Jomini's and Okunev's practical principles, 18; and role of analytical history in theory, 14; on the role of history in theory, 14; and theory of military geography, 12, 14, 15, 16; and theory of strategy, 13, 14, 17, 18, 19; translation of Jomini's work, 13

Zaionchkovskii, A. M., 154, 161

Zeddeler, Baron von, 6, 8, 20, 34; and compilation of the *Military Encyclopedia Lexicon,* 17; and conflict with Sukhozanet, 10; and conflict with Yazykov and Medem, 8; and dismissal from Academy, 10

Zeddeler, L. L. (the younger), 76, 96

Zolotarev, Professor, 117

Zuev, D., 154

About the Author

CARL VAN DYKE is a researcher at the University of Edinburgh, Scotland.

www.ingramcontent.com/pod-product-compliance
Ingram Content Group UK Ltd.
Pitfield, Milton Keynes, MK11 3LW, UK
UKHW020816290926
15470UKWH00003B/1187